Basic Principles of
Environment and Resources

Basic Principles of
Environment and Resources

S.K. Ahluwalia

ABD PUBLISHERS
Jaipur ■ New Delhi

ABD PUBLISHERS

Regd. Off. "Bony Residency", Gate No. 2
Opp. Tilak Public School, Vishweswariya Nagar,
Gopalpura Road, Jaipur - 302018, India
Ph: 0141 - 2761280, Telefax: 2761381
Email: abdpublishersjpr@gmail.com

Delhi Office: 102, 1st Floor, Satyam House,
4327/3, Ansari Road, Darya Ganj,
New Delhi - 110002, India
Ph: 011-45652440

visit us: www.oxfordbookcompany.com

Published in India, 2024

ISBN : 978-8183760147

Price : ₹ 1575/-

Typeset by Shivangi Computer, Jaipur

Printed and bound in India by Thomson Press Ltd., New Delhi

Preface

Human activities generate and exert a great pressure by exhausting nonrenewable resources such as mineral reserves, overexploiting renewable resources such as fisheries and forests, pollution, intruding into ecosystems and degrading basic goods like water and air causing damage to human health and well-being. This results in depletion of natural resources and environmental degradation. The underlying causes lie in the determinants of these individual decisions: the available information on the environmental effects of resource use; preferences of consumers; future effects of current actions; the property rights that define their endowments and set of relative prices of opportunities with these endowments; the cultural, religious and legal restrictions on individual behaviour that prescribed the range of admissible actions and so on.

With growing awareness and importance of environmental and resource problems, efforts have been made worldwide to look into the matter and its seriousness. These have resulted in framing of policies into virtually every sector of social order making requisite understanding of the basics of environmental and resource policy. This is necessary not only for those individuals in policy-making at the local, regional, national and international level, decision-makers in firms and consumer organisations but for students also who have to deal with environmental and resource problems.

"Basic Principals of Environment and Resources" makes investigation into causes, activities, rules and principles affecting natural resources and health of environment. This research takes into consideration effect of markets, external factors, public policies, dynamics, resource utilisation, welfare factors and numerous other activities impacting resources and environment.

S.K. Ahluwalia

Contents

	Preface	*v*
1.	Environment and Resources	1
2.	Markets and External Factors	19
3.	Public Policies and External Factors	55
4.	Dynamics of Environmental Resources	81
5.	Natural Resource Utilization and Environment	131
6.	Welfare Factors and the Environmental Resource Base	167
7.	How to Value the Environment?	187
	Index	*217*

1
Environment and Resources

Introduction

One of the most serious problems the world is facing today is the deterioration of the natural resources. It occurs at any scale, ranging from the individual to the entire world.

The biosphere provides mankind and other species with a place to live and with all material inputs on which livelihoods depend. It thus provides 'environmental utilization space' for present and future use. The biosphere represents our life support system and provides mechanisms for reproducing resources and restoring environmental quality. But these mechanisms are vulnerable, and their 'productivity' is restricted. By generating environmental pressures, human activities adversely affect the size of their own environmental utilization space. In a dynamic sense, these pressures imply the risk that future generations will have less environment to draw their welfare from! Human activities generate such pressure by: (1) exhausting nonrenewable resources such as mineral reserves; (2) overexploiting renewable resources such as fisheries and forests; (3) adversely influencing the rates of regeneration of renewable resource through pollution; (4) intruding into ecosystems; and (5) degrading basic goods such as clean air and water so that damage to human health and well-being results. Economic activity-production and consumption—is primarily undertaken to serve human needs, but the ultimate effects may be destructive in terms of what happens to the welfare of many, now and even more so in the future.

Currently, about 40 per cent of the biological (primary) productive potential of the land (that is, the plant material that the land produces) is now used (and sometimes wasted) by mankind; the tendency is for this percentage to go up to 80 per cent in 2030. By 2050, the world population is expected to have at least doubled. And if this population is to have reasonable standards of living, industrial output may have to be up to ten times the current world production. Exhaustible resources such as oil and several metals are becoming scarce already. Loss of top soil and global warming indicate severe environmental degradation. The implications in terms of costs (that is, agricultural losses) of global desertification have been estimated to be in the range of US$ 26 billion per annum. Forested areas are cut at incredible rates, and the amount of arable land per capita is declining.

This degradation of the environment is the product of the independent decisions of billions of individual users of environmental resources. The underlying causes of environmental degradation accordingly lie in the determinants of those individual decisions: the available information on the environmental effects of resource use; preferences of consumers; the technology available to producers; the rate at which users discount the future effects of current actions; the property rights that define their endowments; the set of relative prices that determine market opportunities associated with those endowments; the cultural, religious and legal restrictions on individual behaviour that prescribed the range of admissible actions, and so on. While the decisions of resource users may be privately rational, they appear to be socially damaging-compromising the interests of present as well as future generations.

Policies to combat these problems are being developed and their effects are felt virtually everywhere. Individuals are households face several serious restrictions on their consumption and behaviour patterns. They come up against annoying problems such as separating waste, and they face an increasing number of new or additional taxes on goods and activities which have a bearing on the environment. Firms are required to reorganize their production processes and to invest heavily in environmentally friendly capital. They also come up against rising input prices and an increasing number of charges on emissions.

National governments are confronted with the problem of deciding upon the optimal level of environmental quality. In this regard a trade-off exists between convential welfare aspects, such as income and employment, and welfare in a broader sense, including environmental quality. The problem is complicated by uncertainty with respect to the physical nature of the environment and natural resources, as well as with respect to economic aspects, such as the preferences of future generations. Additional complications stem from the dynamic aspects of the economic and the ecological systems as well as their interactions. Once a decision about the level of environmental quality has been concluded, the problem of the choice of instrument to achieve the goals fixed arises. Not only national governments but also local and regional governments are confronted with environmental problems. To the extent that it is within their competence, they make trade-offs between environmental quality and traditional aspects of welfare. Moreover, they play a role in the execution of nationally designed or nationally authorized environmental policies.

Environmental and resource problems have become increasingly international in the sense that the consequences are not confined to the country where the problem concerned originates. Examples are the overfishing of oceans, the pollution of transboundary rivers, acid rain and global warming. In the case of international environmental problems there exists no institution (government) with the jurisdiction to enforce environmental policy. Instead, they are based on voluntary cooperation of independent and sovereign states.

This international nature of environmental problems and the need to address environmental problems at the international level have been clear ever since the first report to the Club of Rome: 'Limits to Growth' in 1971. It was further elaborated at the conference on Environment and Development in Stockholm in 1972 and in the First Environmental Action Programme of the (then) European Community in 1973. New dimensions of the international nature were developed by the World Commission on Environment and Development (the so-called 'Brundtland Commission') in the report: 'Our Common Future' (WCED, 1987) and at the second UN global environmental conference on environment and development.

The Brundtland Commission was also the first official commission to adopt the notion of sustainable development as a prerequisite for continued societal existence. It was mainly responsible for putting this notion on the international agenda. Due to the way the notion was interpreted in the report, and enhanced role for environmental economics in actual policy has become a distinct possibility. In fact, the need to look at the economic process as it relates to environmental issues was made clear as early as 1967, when Mishan's *The Cost of Economic Growth* was published.

Critical objectives for environmental and economic policies compatible with sustainability are :

- Reviving economic growth.
- Changing the quality of growth.
- Meeting essential basic needs.
- Ensuring a sustainable level of population.
- Conserving and enhancing the resource base.
- Reorienting technology and managing risk.
- Merging environment and economics in decision-making.

The latter strategic imperative is meant to induce a change in attitudes, objectives and institutional arrangements at all levels ranging from the individual household and firm to the international community of states. The need for this merger emanates from the existence of, and pervasive nature of, environmental externalities, institutional myopia, sectoral self-interest and so on. Sustainability requires changes in the legal and institutional frameworks that will enforce the common interest, that is, environmental protection.

In the process leading up to UNCED, economic issues were seen as crucial. To begin with, the patterns of production and consumption in the industrialized market economies were recognized as major contributors to environmental degradation. Secondly, the right of developing countries and economies in transition to continue to develop was underlined. The two together point to be need to redistribute the access to the planet's limited 'environmental space', and the need to use that space efficiently and sustainably. Thirdly, at various places in the main outputs of the conference there is recognition of the polluter pays principle (PPP) and user pays principle, the need to internalize environmental costs, the precautionary approach to environmental change and the use of economic instruments.

Two decades before UNCED, the growing severity and pervasiveness of pollution in the industrialized economies had led the Organization for Economic Cooperation and Development (OECD) to elaborate, and in 1972 to adopt, the PPP as a background economic principle for environmental policy. The basic tenet of the PPP is that the price of a good or service should fully reflect the total cost of production and consumption. Its aim is to integrate the use of the environment (including its waste assimilation capacity) into the economic sphere through the use of economic instruments such as pollution charges and permits.

Effective international use of the PPP requires a coordinated approach because environmental regulations can become a source of trade distortion if some countries subsidize private investment in pollution control while others do not. This applies in particular to unions of states, such as the European Union (EU). At the European level, the focus of much environmental legislation has shifted towards the European Commission (EC) in Brussels, through the medium of environmental directives. Piecemeal moves to protect the environment up until the Single European Act of 1987 were consolidated under the Act, and environmental protection was given an explicit place in the treaty via Articles 130R, 130S and 130T. The establishment of the Internal Market as from 1 January 1993 required the harmonization of different national laws; environmental measures will also need attention in this context. The goal of harmonized minimum environmental standards is to prevent 'pollution havens' and damage to human health and ecosystems. The PPP was endorsed by the EC in a 1975 recommendation and was included in the Single European Act. The EU in its Fifth Environmental Action Programme is proposing to broaden the range of instruments of environmental policy, including an economic approach of 'getting the prices right'. Targets and actions proposed include the development of indicators of renewable resources, modification of key economic indicators better to reflect environmental effects, application of the PPP, the introduction of a transparent system of pollution charges, deposit/refund systems and so on.

All of the above goes to show how important economic considerations on environmental policy matters are, and vice versa, how important environmentally friendly economic development patterns are.

The growing importance of environmental and resource problems and the increasing infiltration of environmental policy into virtually every sector of the social order require a widespread understanding of the basics of environmental and resource policy. This applies to those individuals involved in policy-making at the local, regional, national and international level, as well as decision-makers in firms and consumer organizations. It also applies to students who somehow will be confronted with environmental and resource problems.

The present volume provides an introduction to the basics of the economic aspects of environmental and resource problems and policy. Following the present introduction, it is made up of three parts. Part I, 'The Environment and Natural Resources in an Economic Pers-pective' focuses on the economic causes of environmental and resource problems and, the other way around, on the economic consequences of environmental and resource degradation. It also deals with principles of public policy. Part II, 'The Economics of Environmental Policies', discusses the goals and principles of environmental policy and gives a comprehensive overview of the various instruments applied in practice. Moreover, ways to analyse the economic implications of environmental policy are discussed. It also deals with the valuation of environ-mental quality, and environmental policy in an inter-national context. Part III, 'The Firm and the Environment' shifts the focus from the public policy-makers to the business policy-makers.

In the next sections the various parts are outlined in greater detail.

The Environment and Natural Resource Utilization in an Economic Perspective

There is a widespread wisdom that markets are the most effective and efficient institutions to achieve socially desirable outcomes. On the other hand, ample evidence exists that the environmental problems we are facing are to a large extent attributable to the role of markets. This does not only apply to the environment in a narrow sense, but also in a broader sense, including natural resources. The purpose of Part I is to analyse what markets can do and what they cannot do, as well as to discuss how markets can be corrected.

'Markets and Externalities', starts by discussing the working of markets. Special attention is paid to market failure caused by externalities and mechanisms to correct for externalities. An externality shows up if one consumer or producer cares directly about another agent's production or consumption, as in the case of pollution by a neighbouring firm. Markets fail in the case of externalities in the sense that goods, such as environmental quality, are not sold on markets. If externalities are present, the market will not necessarily result in an efficient provision of resources (that is, environmental quality). This means that social institutions other than the market, such as the legal system or government intervention, are required to obtain efficiency. The cases with and without the need for intervention are identified. Moreover, some basic forms of government intervention are dealt with. Hence, the foundations are laid for a discussion of the instruments of environmental policy in Part II.

'Public policies and Externalities' focuses on the behaviour of the government from an economic point of view. The main problem is: what environmental problems are regulated by the government? Moreover, attention is paid to government decision rules and the basic principles of the proper level of environmental policy-making. With respect to the former, special attention is paid to direct democracy, including the role of interest groups and bureaucratic influences. With respect to the latter, the discussion focuses on the decentralization principle. Some basic aspects of the choice of instrument are also dealt with.

'Dynamics and Uncertainty' deals with the fact that environmental problems are long-run problems and are cursed with uncertainty. Consider, for example, global warming. Not only are there uncertainties about the link between growing atmospheric concentrations of greenhouse gases and eventual global warming, but also about the magnitude and timing of the warming. Also, uncertainties exist about the economic consequences. Some economists estimate the annual costs of a 2.5^{0}C warming to be more than 3 percent of the gross global product (GGP), while others conclude that the overall impact upon human activity is unlikely to be larger than 2 per cent of total output. It is evident that this complicates decision-making in environmental policy. There is a trade-off in controlling global warming by cutting back carbon dioxide emissions now, and hence incurring GGP losses, versus delaying such actions to the future when more information is available and uncertainty is reduced. However, the risks of the effects of the accumulation of CO_2 and other greenhouse gases in the atmosphere also have to be taken into account. The economic approaches to dynamics and uncertainty are discussed. In the context of the former, special attention is paid to thresholds and irreversibilities, discounting and optimal growth theory, taking into account the interests of future generations. The discussion of uncertainty focuses on modelling uncertainty and decision-making under uncertainty.

'The Economics of Natural Resource Utilization' deals with a topic which is highly important in itself, as optimal resource utilization is a basic aspect of sustainable development. Moreover, non-optimal resource utilization often leads to serious environmental problems. It starts with a classification of natural resources and a discussion of one of their basic features, that is, property rights. Next, attention is paid to nonrenewable resource utilization. The basic utilization model developed by Hotelling is discussed, as well as extensions of this model and taxation of resource extraction. The question of whether or not nonrenewable resources have become scarce is also dealth with. Next, attention is paid to renewable resources. Lastly, two examples of renewable resource utilization, that is, fisheries and forestry, are presented. With respect to the former the debate between maximum sustainable yield and economic optimality is taken up. Next, attention is paid to overfishing in open access fisheries and to various solutions to this problem. In the context of the economics of forestry, the optimal rotation period and forest management planning at the stand level are dealt with.

The 'Welfare Indices and the Environmental Resource Base' deals with national accounts and environmental resources. Although not originally intended, gross domestic product and per capita gross domestic product have come to be almost synonymous with well-being and economic development. However, the present national accounting system ignores the environment. This seriously hampers economic analyses, forecasts and policy-making, as it has become quite obvious in the past few years that environmental quality is a basic component of well-being, besides the conventional indicators. This chapters focuses on the definition of a generalized national product including the environmental and natural resource components.

The Economics of Environmental Policies

In Part II a closer look is taken at how the market failures analysed in Part I could be and, in actual environmental policy are, corrected. The objective of this section is to show how an economic analysis of environmental problems and practical approaches to them, may assist in achieving acceptable and workable environmental policies. This implies that we look in detail at what environmental policy wishes to accomplish and in what circumstances it operates. This is to be put in an economic context: (1) what is the economic rationale for embarking on such a policy venture; and (2) how will environmental policies and instruments affect the economic process? Moreover, methods to analyse and evaluate the economic costs, as well as benefits, of environmental policy interventions are dealt with.

The focus is on criteria such as effectiveness, efficiency, equity and sustainability, and on policy principles such as the PPP, 'polluters pay', 'users pay' and 'victims pay'; these serve to allocate the costs of achieving policy objectives. Attention is paid to typical 'policy failures'—as economists might help identify. The notion of 'optimal level of pollution' is discussed (which may be a higher level of pollution than environmentalists would aim at) as well as 'perverse' incentives that the market mechanism may emanate towards the economic agents. The conclusion is that economic analysis may assist policy-makers in assessing costs and benefits of alternative interventions and in the development of appropriate tools or instruments for environmental policy.

Of course, the question arises of what the benefits of policies protecting the environment are, and this leads to a discussion ('Valuing the Environment') of various techniques of valuing environmental goods and services. Here the point of departure is a 'welfare theoretical' underpinning of environmental functions. Several techniques of attaching monetary values to these functions are discussed, that is, direct valuation, the travel cost method, hedonic pricing and contingent valuation. The scope of these methods is to capture the societal significance of environmental quality. It is noted that policy-makers make little use of these techniques, sometimes for understandable reason, but often at a cost in terms of suboptimal decision analysis.

Environmental policies cannot operate effectively without instruments. 'Environmental Policy and Policy Instruments' discusses and compares several categories of tools: regulations and standards, environmental taxes and charges, tradeable permits, deposit refund systems, damage liability and compensation. We see all of these in operaton, though not in all countries. The focus here is on experiences in OECD countries. It appears that different groups of countries employ different sets of instruments to address basically the same problems because of different conditions, for example, socio-political settings. Nevertheless, countries may learn from one another's experiences, and environmental economics may provide useful insights.

The environmental policies of countries will affect the economic performance of the various agents. This may be especially important when there is a possibility that the competitiveness of firms in different countries may be affected differently with implications for international trade and the international distribution of welfare. Hence there may be a need to coordinate the policy approaches of the various countries in order to reduce or avoid distortions.

Another rationale for an international approach is that pollutants may cross national borders, as in the case of marine pollution, acidification and so on. 'Inter-national Coordination of Environmental Policies'. The gains of coordination are made visible by reference to the example of anti-acidification policies. Also the trade effects of environmental policies are looked into and recommendations are derived as to when and where to harmonize.

Having established that achieving certain environmental objectives through policy interventions may have benefits and what these interventions may be the question naturally arises as to the costs of environmental policies in social or macroeconomic terms. Calculating these is of great policy relevance, but by no means an easy and straightforward exercise.

'Modelling the Economic Consequences of Environmental Policies' discusses the main types of economic models to approximate these costs, and their main adva' 'ages and disadvantages. Examples are given for eacn type of approach, mainly in the area of energy /carbon policies. The models discussed are: partial equilibrium models, input-output models, ordinary macroeconomic models and the applied general equilibrium models. It turns out that these models may be useful for different purposes and time horizons, and imply different data requirements and assumptions about the economic process.

'Environmental Policy in the European Union' discusses the development of environmental policies in the EU. The EU policy on the environment has changed over time to a clearly distinguishable, separate policy concern at the EU level. An important set of practical principles governs the domain and scope of EU environmental policies. For example, the 'subsidiarity principle' demarcates the national responsibilities from those of the Union. The current Fifth Environmental Action Programme is discussed with special attention to the instruments used or discussed. As a case study, details are provided on proposals for an EU on carbon emission reductions and how this policy progresses.

The Firm and the Environment

Part III, 'The Firm and the Environment', shifts the focus of the volume from the public policy-maker in Part II to the business policy-maker. It is important to pay explicit attention to the firm for several reasons. First, firms are rarely passive in the public policy-making process. They are usually active participants-and sometimes even the protagonists - in virtually every story of policy formulation and implementation. 'Environmental Policy in the European Union', shows this vividly in the case of the European carbon tax proposal which firms effectively sabotage. So it is important that we examine the role firms play to see to what extent (and why) they are supportive or injurious to the implementation of the policy ideas discussed in this book. The first two chapters of Part III deal with these questions directly. Secondly, in market economies, managers of private sector firms make the actual allocation decisions for environmental resources as far as production is concerned. If the public policy measures discussed in Part II are to succeed in mitigating the problem of market failure covered in Part I, they will do so only indirectly by inducing an improvement in the quality of the decisions made by these managers. So we must trace the path of environmental policy through the firm to its ultimate consequences for the efficient allocation of environmental resources.

Finally, environmental regulations and public pressure for environmental protection have stimulated the development of one entire industry-that of waste management. Its effectiveness will depend critically on how firms are structured and how they compete in both the regulatory and the market arena.

'A Firm's Involvement in the Policy-making Process', investigates the circumstances in which a firm's involvement within the regulatory process takes place and the pay-offs provided by this involvement. After presenting both empirical and theoretical arguments to sustain the claim for the involvement of the firm, the author looks explicitly at four forms of such involvement. These forms include support for public policy, its obstruction, voluntary action under regulatory threat, and cooperation with the government to design environmental policy.

'Environmental Management as a Competitive Strategy', presents the case of chlorofluorocarbons, or CFCs. The case illustrates a 20-year evolution in firms' thinking about environmental issues and their strategic significance. It also covers an evolution in public policy thinking, and the reader will see how evolving corporate and public policy strategies become intermingled in what turns into a dynamic game. The reader will see how firms compete with one another on the environmental dimension in both the market-place and the regulatory arena, and how countries compete as well. The case gives one a chance to question, in principle and in practice, the role of private business firms in influencing public policy formulation.

It is customary for both environmental economists and policy-makers to assume that the firm is a 'black box', within which 'rational' managers make profit-maximizing decisions that we can be modelled without any need to 'pierce the corporate veil'. Such an assumption ignores many issues that are intrinsically interesting and relevant to the actual success of public policy. The reader will see in 'Corporate Responses to Environmental Concerns', that firms are not perfectly efficient black boxes, and that they suffer from organizational failures which are analogous in many ways to the market failures covered in other parts of this book. Managers must try to correct organizational failures just as public policy-makers must try to correct market failures, and neither will ever succeed perfectly. Systems design in both areas must take this reality into account.

One cause of organizational failure is imperfect information. 'Environmental Information Systems and Eco-auditing', explicitly addresses the management and the public policy challenge of improving the quality of this information and its use. As the author says, what is needed-in the hands of managers as well as stakeholders, who have the power to influence the public image of a company and its economic welfare - are comprehensive environmental information systems, which give a reliable picture of the firm's relations to society and nature. The author goes on to discuss the basic principles of eco-accounting and eco-auditing and to describe several systems employed by firms in different countries.

One objective of any environmental policy is to encourage innovation. Indeed, the impact of regulation on innovation may in the long run be of greater importance than its impact on static resource allocation efficiency. 'Environmental Technology', considers the incentives for, and the means by which, environmental technology is developed and adopted by industry. It starts with an exploration of the environmental technology concept and then identifies relevant insights from the economics of technical change. The author examines the evolution of the market for environmental goods and services, and the role which government activities have played in it. The chapter ends by drawing together key conclusions regarding the factors influencing the development of environmental technology.

Illustrate the implications of environmental issues on the conduct of business affairs in two different ways. The first chapter is concerned with one function in a firm: its operations strategy. The second is concerned with one industry in the industrial landscape, that of waste management.

'Environmental Issues and Operations Strategy', considers a particular domain within a firm and examines how it is influenced. The term 'operations' traditionally refers to all aspects of product and process design, production and distribution logistics. More and more, it includes reuse, recycling, remanufacturing and reverse logistics (reverse materials flows). Using a well-known framework for operations strategy, the authors look at many operations variables including, among other things, cost, quality, reliability, flexibility, capacity planning, facility location, inventory control and capital allocation. They conclude that implementing environmental programmes in the operations function requires no radical overhaul of operations strategy because of the analogies the authors find to existing management concepts used there.

'Waste Management', is about the particular and important problem of solid waste and the industrial system which must be designed to deal with this problem. Developed economies generate quantities of waste that exceed the environment's absorptive capacity, requiring decisions regarding policy emphasis on source reduction, recycling and/or alternative means of final disposal of many different kinds of waste. The reader will find in this chapter references to many of the issues raised earlier in the handbook—market distortions, policy principles and problems of organizational failure, as the author argues his point that economies requires, but currently lack, an integrated, effective and efficient waste management system.

Conclusions

Having outlined, in sections 2, 3 and 4 above, the contents of this volume, we now return to some of the observations presented in section 1. We ask ourselves to what degree environmental economics can indeed help in achieving sustainable development or in keeping the economic process within the constraints of a bounded environmental utilization space.

Economic activities have been identified as the principal 'carriers' of environmental pressure, in the form of pollution, resource demands and spatial claims. Many of the life support functions of the biosphere are now at risk and, locally or regionally, economic development has become unsustainable. Driving forces at work which should be taken into account in this regard are technological innovation; population growth, the poverty gap between the various regions of the world and economic growth tendencies. This volume has, though not fundamentally, addressed these driving forces or ways to curb them. Moreover, this volume has not looked deeply into the institutional and cultural settings within which these forces operate. Several of the themes that have emanated from, for example, the UNCED conference such as the need to search for alternative patterns of production and consumption, or the need to develop techniques for managing the global commons, have been left aside. We have, however, picked up a major methodology to address these issues and to contribute to sustainable development, that is, 'merging environment and economics in decision-making'. This has been addressed both at the level of public (national and international) and private policy-making (firm level).

We have made clear how environmental degradation can emanate from economic activity in the form of, as economists call them, 'externalities'. And we have pointed at a number of institutional factors that explain their coming about, such as market failures and policy failures. In order to curb these failures most economic institutions and policies need to be fundamentally restructured. Various options have been identified that could help in 'ecologizing' the economic processes on its way to sustainability. Property rights have to be redefined and new ones must be put in place. Liability and product chain responsibility must be (re) formed, as well as a range of tools or instruments that societies can deploy to harness the market mechanism in order to make it operate in a socially and environmentally more appropriate way. Information systems for environmental change at the national, regional and firm level, and methods to reflect such change in the System of National Accounts, must be developed as well as the tools to assess the economic significance of environmental policy interventions.

Europe is facing difficult economic times. Large parts of Europe are in the process of economic restructuring and even of systematic transition. The poor economic performance of many European economics is eroding the societal support for measures aimed at achieving sustainability. In the 1980s, the advanced industrialized economies went through a process of 'decoupling' of environmental pressure from economic growth: as income per capita rose, environmental impacts went down, both in relative and in absolute terms. In the area of many pollutants this still appears to be the case. However, when it comes to environmental inputs, some early warning signs indicate that we are now in an upswing again: environmental pressure and economic growth may be 'recoupled'. Employment issues feature at the top of the social agenda, and for good reason. In 1993 the former President of the European Commission, Delors, published the 'White Book towards the 21st century: Ways and Challenges'. In it, views are presented on a new European model of development, and attention is paid to the structural links between environment and employment. Labour is currently underutilized whereas environmental resources are over-utilized. A new model for 'sustainable development' in which this dilemma is addressed is advocated. In this way economic development and environmental quality can be reconciled to enhance societal support for environmental policy.

Clean technology is identified as a key factor, and so are methods to facilitate the necessary structural changes: (1) a strategic microeconomic policy such as 'getting prices right' and providing well-targeted incentives; (2) appropriate macro-economic policy instruments such as environmental taxes, fiscal facilities and international cooperation. An environmental infrastructure (recycling, waste treatment facilities, and so on) is to be developed, and a carbon/energy tax might pave the way for a tax system that lowers labour costs and reduces environmental pressure. Most of these ideas are in fact discussed in this volume from an environmental economic perspective. As we suggested at the beginning of this section, such an approach may indeed basically only curb environment-economy antagonisms rather than address them head-on, but at the same time it provides ample material that-if put into practice-would move Europe forward on the way to sustainable development.

2

Markets and External Factors

Introduction

There is conventional wisdom that market economies mostly lead to socially desirable outcomes. Yet it is also a well-known fact that economic activity frequently has socially undesirable environmental consequences. Experience tells us that things are not always what they seem, 'a paradox is a truth that stands on its head to attract attention'. The purpose of this chapter is to discuss, intuitively, what markets can do-and what they cannot do—in terms of allocating resources in a socially desirable manner. In this way, we can reconcile the seemingly conflicting outcomes of economic interaction.

We will also consider possible remedies to what economists call market failures. We will concentrate exclusively on failures which are caused by what we refer to as externalities. This concept will be defined more fully below, but for the moment the reader may consider it the unwanted environmental consequences of economic activities.

In a market economy where externalities are present, there are, theoretically, many apparently similar ways in which environmental policy can improve the allocation of resources. However, in practice, it is not clear which policy measures are the most successful. Therefore this chapter also contains a discussion of the pros and cons of different practical policy applications.

The Blessing of the Market

Generally, you do not buy goods that do not improve your personal situation. At the same time, you would not be able to buy if the seller felt that he would be the loser in the exchange. Hence given any initial allocation of goods among traders, the opening up of trade cannot be but an improvement of the initial allocation. A trader would always block any disadvantageous trade offer. Moreover, if a group of individuals wants to trade, it is not inconceivable that they will find all the mutually advantageous trades, sooner or later. If the number of individuals is large this will be difficult, or at least take a long time, when there are no coordinating (market) arrangements. The human brain would soon, however, create tools to simplify the discovery of advantageous trades. Say, for example, that a market administrator is chosen who starts the trading process by announcing an arbitrarily given set of prices. Agents specify what they are willing to buy and sell at the prices quoted. If the buying and selling decisions are inconsistent, the market administrator adjusts the set of prices according to the following rule: if, in a particular market, the aggregate quantity demanded exceeds the aggregate quantity supplied, he will raise the price in that particular market; if the opposite holds, he will lower the price. The agents are allowed to renew their bids, and the process described above, called the *tatonnement* process, continues until a simultaneous consistency is achieved of buying and selling decisions in all markets. Then actual trading is taking place.

What are the conditions under which trading takes place? Obviously, since everybody pays (gets) the same price and buys (sells) all that he wants at that price, the utility of all individuals cannot be improved by a reallocation of goods from one individual to another. If we give any demander one more unit this will improve his position, but the position of the supplier will deteriorate, since he already trades all that he is willing to trade under the ruling trade conditions. To see this more clearly consider Figure 2.1.

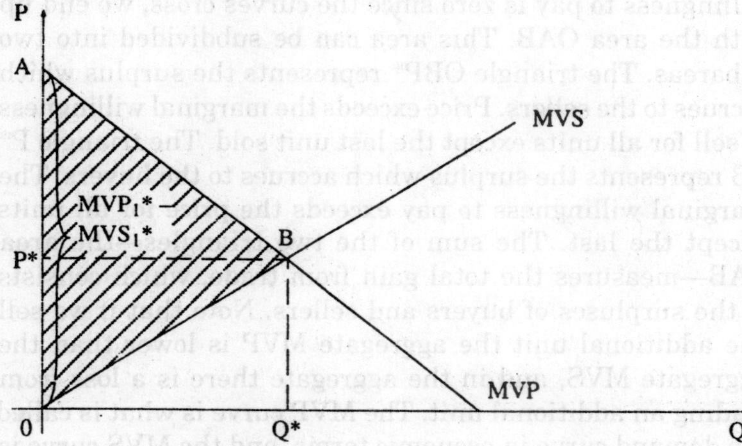

Fig. 2.1 : The totem of the market or the maximization of the surplus.

The 'marginal willingness to pay' curve, MVP, denotes the aggregate net demand of good Q at any given common price (measured along the vertical axis). The word 'marginal' is used to highlight the fact that the MVP curve tells us what the market is willing to pay for the last unit bought. It is evident that the higher the price, the lower the aggregate demand. Similarly, the 'marginal willingness to see' curve, MVS, represents the aggregate net supply of good Q at any common price. It gives us the price which the market accepts to sell the last unit. The price p*, at which the MVP curve cuts the MVS curve, is particularly interesting. At this point the aggregate volume sold coincides with the aggregate volume supplied, and all desirable transactions can be carried out. This is an equilibrium and the situation where the market administrator has done his job and trading can take place.

What are the gains from trading? For the first unit, the aggregate marginal willingness to pay exceeds the aggregate compensation the sellers need to supply it on the market by the distance MVP_1—MVS_1. By moving unit by unit towards the equilibrium volume and summing the net willingness to pay for each unit up to Q*, where the net willingness to pay is zero since the curves cross, we end up with the area OAB. This area can be subdivided into two subareas. The triangle OBP* represents the surplus which accrues to the sellers. Price exceeds the marginal willingness to sell for all units except the last unit sold. The triangle P* AB represents the surplus which accrues to the buyers. The marginal willingness to pay exceeds the price for all units except the last. The sum of the two triangles—the area OAB—measures the total gain from trade, which consists of the surpluses of buyers and sellers. Note that if we sell one additional unit the aggregate MVP is lower than the aggregate MVS, and in the aggregate there is a loss from trading an additional unit. The MVP curve is what is called the demand curve in economic terms, and the MVS curve is the supply curve. The point where demand and supply are equal is called the market equilibrium.

It is, of course, true that few real markets have a market administrator or an auctioneer, but economic theory assumes, as a first approximation, that a market where many traders are present, functions as if there was a market administrator (that is, as in a perfect market). The bottom line is, therefore, that the existence of perfect markets guarantees that all mutually advantageous trades are realized and that the outcome is socially desirable in the sense that one cannot improve the situation for one individual without worsening it for someone else. This is essentially Adam Smith's famous 'invisible hand' theorem. A selfish price-taker competing with other intends only his own gain, and he is in this, as in many other cases, led by an invisible hand to promote an end which was no part of his intension,[2] namely an efficient exchange.

Market Failures Caused By Externalities

We have already mentioned that environmental problems cannot normally be handled in a socially desirable manner without policy interactions in the market system.

Given that we espouse a belief in the blessing of the market, we would blame environmental problems on a failure of markets. We will leave this conjecture open for the moment, and turn to a definition of externalities. This will enable us to embed environmental problems in a broader class of phenomena which cause market failures.

Baumol and Oates state that an externality is present whenever some individual's (say A's) utility or production relationships include real (that is, non-monetary) variables, whose values are chosen by others (persons corporations, governments) without particular attention to the effects on A's welfare.

The consumption of antibiotics, for example, causes bacteria to become resistant, which will complicate future recovery from bacterial attack. Secondly, the production of knowledge in society probably affects the production technology of individual firms; the emissions from smoke stacks have a deteriorating effect on the 'living conditions' of both firms and individuals. These phenomena belong to the class of externalities. The examples show that externalities can be both good and bad. The externality created by the production of knowledge is a public good, since it improves the production technology of firms and also the utility of consumers. The consumption of antibiotics contains an element of a negative externality (a bad), since it decreases the utility of other individuals' future use of the drug. It is, however, obvious that the fact that antibiotics cause externalities does not mean that they should not be used; the trick is to balance the harm done against the obvious beneficial effects of the drug in such a way that a socially desirable consumption level is generated. There are few reasons to believe that the demand and supply of antibiotics in a free market would generate the optimal consumption level. This is also recognized by governments: therefore the prescription of antibiotics is regulated (in most countries).

However, the fact that we have to rely on bakeries for our daily bread does not mean that an externality is involved. The baker does not decide for us how much bread we should consume. Likewise, there is no externality involved if an increase in the demand for tankers raises the price of steel and hence affects the welfare of the purchasers of cars. The reason is that there is no direct link between the technology of car production and the demand for steel, that is, the number of tankers sold does not directly affect the technology used in car production. The rise in the price of steel is a correct reflection of the fact that steel has become a scarcer resource, which has pecuniary repercussions for all users—direct or indirect—of steel.

Public versus Private Externalities

Many of the externalities above have the character of public goods, that is, they are consumed by all, and the consumption by one individual does not affect what is left for consumption by other individuals. Broadly speaking, a public goods externality is consumed by all people in equal amounts; for example, air pollution, in terms of sulphur dioxide or carbon dioxide, has an equally deteriorating effect on the environment for every resident in the area. An increase in the population of the area will not reduce the level of sulphur dioxide.

An amenity is another typical example of a public goods externality, but this time a positive one. For example, beautiful views can be enjoyed by all people passing by, providing benefits for all viewers. After you have seen the Grand Canyon it is still there to be enjoyed by later visitors. It is not difficult to understand why it is inefficient to charge for the consumption of positive public goods externalities. This is because the consumption of amenities by one individual does not, in general, influence the level of satisfaction of another individual. Hence a charge would inhibit one person's consumption and satisfaction without increasing that of others. Accordingly, public goods externalities are sometimes referred to as 'undepletable externalities'.

An example of a positive private externality, let it be a little special, is the slaughter of cattle in Argentina around the turn of the 19th century. This was done because of interest in the hides. The meat was left at the place of slaughter to be enjoyed by the poor. This externality no longer exists as the meat is more valuable (nowadays than the hide). It is obvious that this is a depletable or private externality, since a piece of meat taken away by one person will not be available to others. Since it is possible exclusively to charge the person who is enjoying the externality, the externality is depletable. The generator of the externality may find it profitable to produce the good 'with particular attention to the effects on A's welfare'. This means, by definition, that the externality is eliminated.

Another example of a private externality, this time a bad, is a firm that dumps its solid waste outside your summer cottage. The externality is obviously divisible since it cannot destroy the environment surrounding, say, your cousin's cottage in another country.

There are cases where it is not clear whether we are dealing with a typical public or private externality. However, the policy problem tends to be the same for any form of externality: how do we design the incentive structure facing the parties to an externality so as to induce socially desirable behaviour? As it turns out, policy measures for correcting the inefficiencies created by any kind of externality, public or private, are essentially the same. The difference is, perhaps, that the more private the externality, the more likely it is that it can be (will be) regulated by markets and market transactions.

Correcting For Externalities

We will deal with the problem of correcting for externalities in the simplest possible setting. First, we will discuss a case where there are two firms, and where the production of firm 1 causes firm 2 an externality, which is in relation to the production of firm 1. One unit of production in firm 1 creates an externality (a cost) for firm 2 which in total amounts to kq, where k is a constant and q is the number of units produced in firm 1. More particularly, every q units of output in firm 1 creates a loss of kq dollars for firm 2. To keep the analysis as simple as possible, we will not deal explicitly with the production decision of firm 2. Suffice it to say that its profit is:

$$\pi_2 = {}^{-}kq \qquad ....(1)$$

which also shows that every extra (marginal) unit produced in firm 1 inflicts a loss on firm 2 of k dollars.

Firm 1 produces and sells it output in a competitive market, where there are many sellers and buyers. The participants in the market consider, therefore, the market price p as a given constant. The cost of production increases with q. We can write the profit of firm 1 as:

$$\pi_1 = pq - TC(q) \qquad ....(2)$$

The first term is the total revenues, which is the price times the volume sold or the marginal valuation of the good times the volume. The second term is the total cost, and note that the cost firm 1 inflicts on firm 2 is not in the cost function *TC(q)* of firm 1.

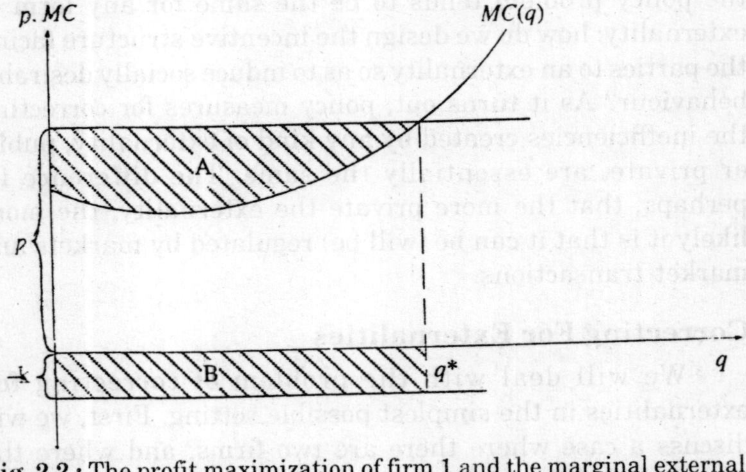

Fig. 2.2 : The profit maximization of firm 1 and the marginal external effects it causes firm 2.

Both firms are assumed to maximize their own profits, although we are not dealing with the maximization decision of firm 2 explicity. How would the production of firm 1 be determined under profit maximization? For each unit it sells, it gets p dollars, and each extra unit costs, *say MC(q)*. Clearly, it would be willing to supply an extra unit as long as the price exceeds the extra (marginal) cost $MC(q)^3$. In Figure 2.2 the revenue relevant for profit maximization and the cost structure of firm 1 is depicted. We have also portrayed the marginal cost inflicted on firm 2 by each extra unit of production in firm 1.

It is obvious that firm 1 makes a profit on all units up to q^*. The profit it has earned up to q^* is marked by the shaded area A. This is also the maximum profit, since if the firm chooses to sell units beyond q*, it will lose on each extra unit, since the marginal cost, *MC(q),* is larger than the price, p, for all units to the right of q*. The value of the externality it causes for firm 2 is the shaded area below the q axis, which is denoted B. It is obvious from Figure 2.2 that it is impossible to avoid the external effect without shutting down firm 1. This is, however, not necessarily the socially desirable outcome. What society would like to do is to maximize the social surplus from the production of firm 1. Here this corresponds to choosing q in such a manner that the sum of its own profits, minus the costs it inflicts on firm 2, is maximized.

Internalizing the Externality

This example leads us to the first method of correcting for the externality. If firm 1 owned firm 2, it would have reason to care about the detrimental effects the production of firm 1 had on firm 2. The decision problem of the owner of both firms would be to maximize.

$$\pi_1 = pq - TC(q) - kq. \qquad ....(3)$$

The revenue function is the same, but the total cost function and the marginal cost function have extra components, kq and k, respectively. We say the externality kq has been internalized by introducing a single owner of both firms. The determination of the socially desirable production level is illustrated in Figure 2.3. The relevant marginal cost function is now MC(q)+k, and the firm makes a profit as long as the price exceeds the augmented marginal cost function MC(q) + k.

The socially desirable level of production is given by, where society's marginal valuation of the good, p, coincides with its social marginal cost, MC = k. We can also see from Figure 2.3 that the social surplus (the profit at the socially desirable level of production) is given by the magnitude of the shaded area A^S and that $0 < q^s < q^*$, that is, the socially desirable production level is positive, but less than the original decentralized solution q^*. It would have been zero if k had been large enough to neutralize all surplus A^S.

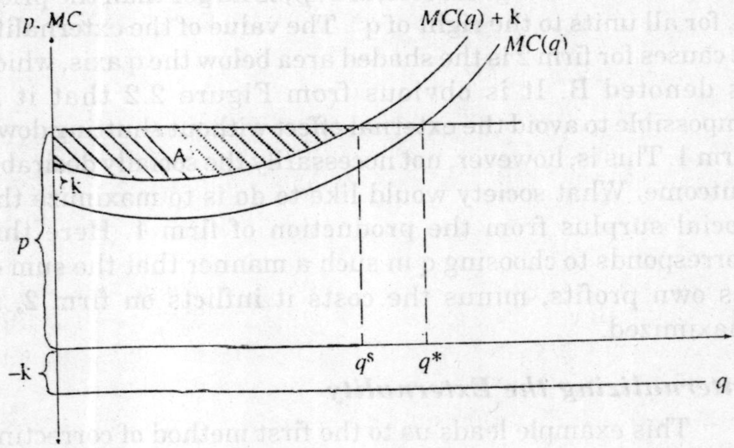

Fig. 2.3 : Profit maximization when the externality is internalized.

If firm 1 owned firm 2 the externality would disappear, since private and social marginal costs would under such conditions, disappear. It is therefore relevant to ask whether there is any incentive for a single owner to materialize. Clearly, the profit (value) of firm 2 increases when production is reduced from q^* to q^s, while the profit value of firm 1 decreases. The aggregate profit, however, would increase after merging of the firms and consequently the production of the socially desirable production level [from A - B to A^s].

This means that there is an incentive for one owner to buy out the other, given that the value of the two firms corresponds to their profits in the free market solution. Hence, given perfect asset markets, the arbitrage possibilities (possibilities of making a safe profit) of a merger would, in theory, sooner or later be taken care of by some party, and the social inefficiency arising from the externality would disappear. In other words, more markets and complete information could automatically correct for the externality.

Pigouvian Taxes

Now we understand how the externality can be internalized, the next step is to come up with an alternative policy proposal to correct the externality.

If we tax firm 1 with a tax *t* per unit equal to the value of the marginal external effect k, the firm will be induced, via the profit motive, to choose the socially desirable level of production. It will stop the expansion of production, where:

$$p = MC(q^s) + t = MC(q^s) + \text{k}, \qquad \text{....(4)}$$

which, from the last equality, results in the socially desirable production level q^s, depicted in Figure 2.3. The tax *t* is called a Pigouvian tax after Arthur C. Pigou, who in 1920 suggested the technique of using taxes (subsidies) to correct for socially undesirable (desirable) outcomes caused by externalities.

This method has long been, and still is, the most widely suggested cure for externalities, at least in textbooks on environmental economics. However, alternatives such as tradeable emission permits (see section 4.4) are growing in popularity.

From the example above it seems as if a direct regulation, given perfect foresight, would do equally well. That is, suppose we told the firm in Figure 2.3 that the maximum allowable output was q^s, would be then not achieve the same allocation as under Pigouvian tax equal to t = k? In the example above the answer is yes, but in a more realistic setting where there are many firms (industry) and in which everyone produces public 'bads', we would like to obtain a given level of cleaning up at the lowest cost in the aggregate. This is very difficult if we do not have exact knowledge of the production technologies at the firm level. It may easily happen that a regulation requiring, say, a 20 per cent cut in emissions from each firm, results in an excessive social cost for achieving a 20 per cent cut in the aggregate emission level. The same Pigouvian tax on each unit emitted, the level being determined by the value of the marginal externality at the social optimum, means that each firm will abate its emissions (essentially 'eliminate' k in the example above) as long as the tax (t = k)exceeds the marginal abatement costs. If the abatement costs exceed the tax, it pays to pay the tax and, if the opposite is true, it pays to abate.

Each firm would then, in equilibrium, have an abatement level determined by the equality between the Pigouvian tax and the abatement cost of the marginal unit (the marginal abatement costs). Since the tax is equal for all firms, they would all have the same marginal abatement costs. This is indeed a necessary condition for a cost-effective abatement policy. To see this, say that the marginal abatement cost differs between firms. It would then be cost saving to reduce abatement in firms where marginal abatement costs are high and increase abatement where costs are low. Such reallocations are not profitable if marginal abatement costs are equal at all emission sources. Clearly, if abatement was determined by a direct regulation, it would only be by sheer luck that regulations were allocated in such a manner that marginal abatement costs were equal for each firm in the industry.

There is at least one more advantage to a tax, as opposed to direct controls, in that the emission tax provides a continuing incentive for the polluter to abate emissions. When a tax is imposed firms are encouraged to seek new low-cost ways of abating emissions, regardless of how much they are already abating, since such a technology would exempt the firm from taxation.

A potential disadvantage of a Pigouvian tax is that extensive monitoring of missions is required to make sure that the desired level of pollution is obtained. This is, of course, also true for individually designed regulations. However, it can be argued that if all polluters are required to use the same cleaning technology, such as a catalytic converter for cars, the administrative costs of the regulation can be kept to a minimum. Hence, if cost-effectiveness and the administrative costs of monitoring are taken together, it is conceivable that there are cases in which a mandated technology will represent the lowest-cost alternative.

There are other obvious situations where direct controls are preferable to a tax. For example, a simple ban is more effective than a Pigouvian tax when dealing with a highly toxic substance. Another situation is where quick alterations in the level of emissions are required to meet, for example, changing weather conditions. Naturally, a tax could be altered to meet varying emission targets, but it is clearly impractical and expensive to do so frequently.

Economists tend to claim that, with a few obvious exceptions, a tax to regulate emissions is generally superior to direct controls. The principle that the polluter pays' has, however, been challenged by economists claiming that a subsidy will work just as well, that is, the polluter is paid to reduce pollution. Our intuition clearly tells us that whether the polluter is paid for the emissions he abates or is taxed for those he does not, the outcome would be the same. This intuition is, however, not quite correct. A subsidy, contrary to a tax, approves the profit conditions for the firms within the industry and, although emissions from individual firms would decline, this is not necessarily true at the level of the industry as a whole. This is because firms enter the market as long as profit are higher than normal and a subsidy means that profits remain higher than normal, for a greater number of firms in the industry than under a tax regime. Hence, too many firms would, in a dynamic perspective, be attracted as the industry, and the allocative effects of a subsidy would not be the same as those of a Pigouvian tax.

In practice, there is an additional difficulty with a subsidy. The level of the subsidy has to be based on some benchmark emissions level, which would give the incumbents an incentive to misallocate resources, emitting more than necessary to establish a favourable benchmark-too favourable from society's point of view.

The conclusion is that whereas our intuition tells us that a subsidy would also be cost-effective, further reflection tells us that a subsidy is not sufficient for a socially desirable outcome.

Another somewhat counterintuitive proposition is that competitive polluting outputs are not necessarily excessive. To be more precise, if there was an externality-producing branch of industry, say the transport sector, out intuition and Figures 2.2 and 2.3 would tell us that an unregulated competitive market solution would generate too many transport services. Broadly speaking, since all activities within the transport sector, transportation by car, train, boat and air, all generate externalities in terms of unwanted emissions, a socially desirable outcome will involve a reduction in all these activities. Again, this intuition is not correct. The picture given by Figures 2.2. and 2.3 is a partial equilibrium view. When all repercussions of appropriate Pigouvian taxes on transport services are taken into account, the final outcome could very well mean less car and air transportation, but more boat and railway transportation. The reason as fairly obvious. The emission caused by the train per ton/mile is less than that caused by the car. It is clear, then, that a socially desirable outcome may call for a decrease in the use of automobiles (compared with the unregulated market solution) and some offsetting increase in the use of railroads, in spite of the fact that trains produce undesirable emissions. Total emissions would certainly decrease as a result of the suggested substitutions.

Another much debated issue among economists is whether the victims of externalities should be compensated. If the number of victims is large, efficient treatment of them prohibits compensation, and this is true for both depletable (private) and non-depletable (public) externalities.

The reason is related to the fact that taxes are preferable to subsidies as a means of controlling emissions. If all the neighbours of a factory were fully compensated for, say, the effects of smoke, no one would have any incentive to move away from the factory. It is then certainly conceivable that too many people would choose to live under smoky conditions and be fully compensated - without benefit to anyone. Broadly speaking, too many people would engage in 'victim activity'.

This point can be enlarged upon. It could be argued that the victims should be taxed for the cost their decision to locate near the polluting factory imposes on the factory owners. Coase's idea is that when residents select a home near the factory they impose an external cost on the generator of the externality. This cost takes the form of a higher Pigouvian tax for the owners of the factory, reflecting greater social damage from the pollution due to the rise in the number of victims. However, if the number of victims is large, the Pigouvian tax will essentially be independent of the number of victims and one can neglect this social cost from locating close to the factory. The conclusion is that correct incentives should be provided to victims so that they take defensive actions, such as investing in air-cleaning devices or moving to new locations further away from the source of pollution, that is they should neither be compensated nor taxed.

Markets and Property Rights— The Coase Theorem

Ronald Coase has also presented an interesting view on how direct market transactions between polluter and victim can, without any interventions from policy-makers, produce socially desirable outcomes. The setting is an economy where property rights are well defined where there are few polluters and victims. We will illustrate Coase's idea with an example from a conflict over land use.

In the northern part of Scandinavia there is a conflict over land use between the industries of reindeer husbandry and forestry. Both industries use common land. On the one hand, the Laplanders (the reindeer-owners) claim that modern forestry destroys, among other things, the pastures for reindeer. On the other, there are the claims of the forestry industry that the reindeer destroy the plants during the regeneration period. The conflict over land use is a clear example of external diseconomies. The production function of forestry depends on the intensity of reindeer husbandry, and the production function of reindeer husbandary depends on the intensity of forestry. To simplify the analysis, we assume that the intensity of forestry affects reindeer husbandry negatively, while the effects of reindeer husbandry on forestry are negligible.

Let $B^F(x)$ be the net revenue from forestry, when forestry is conducted at intensity x. Moreover, let $B^R(x)$be the net revenue that falls on reindeer husbandry when forestry is conducted at intensity x. The socially optimal scale of forestry is the solution to:

$$\max_{x} S(x) = B^F(x) + B^R(x), \qquad \text{....(5)}$$

that is, to choose x, the intensity of forestry, such that the sum of the net revenues from forestry and reindeer husbandry is maximized.

The same intuition that helped us derive the conditions for cost-effectiveness now tells us that x must be chosen such that the net revenue of the last unit (the marginal benefit from forestry) equals the marginal net cost it imposes on reindeer husbandry: the marginal benefits of forestry should be equal to the marginal cost forestry imposes on reindeer husbandry.

The solution is illustrated in Figure 2.4. At the optimum x^* the net revenue from the last unit produced in forestry, $\frac{dB^F_*}{dx}$, equals the net cost which falls on reindeer husbandry,

$-\frac{dB^R_*}{dx}$. If forestry were allowed to choose its own harvest rate, it would harvest until marginal net revenue is zero, that is, x^F was produced. Thus we see that the private optimum results in a level of forestry activity which is too large. On the other hand, if reindeer husbandry were allowed to decide the rate of forestry production, forestry production would not be allowed at all, that is, point x^R would be chosen.

This conflict over land use may be solved by internalizing the externality through allowing forestry also to conduct reindeer husbandry. The goal functions of society and forestry in such circumstances would coincide. Other possible ways of solving the conflict are direct regulation or the use of taxes in the way discussed in section 4.2.

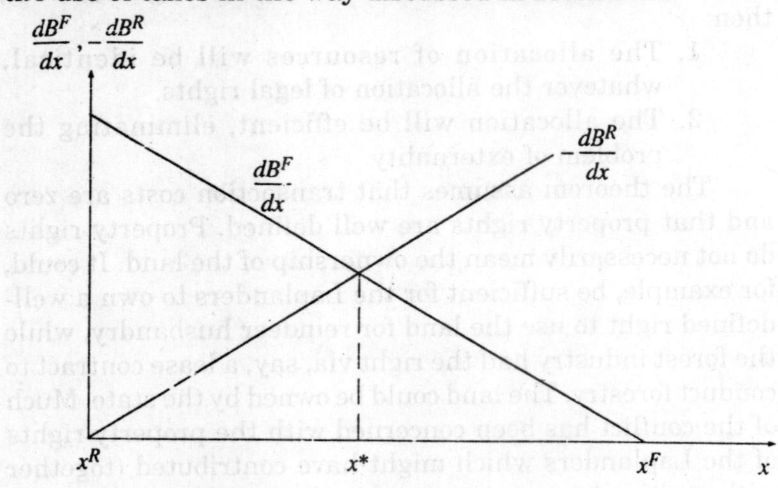

Fig. 2.4 : Illustration of Coase's theorem.

But why should the reindeer husbandry industry allow a level of forestry production which is too high? Both forestry and reindeer husbandry would gain if forestry were compensated for decreasing its harvest rate by a sum of money (ε, where $0 < \frac{dB^F}{dx} < \varepsilon < -\frac{dB^R}{dx}$) as long as the intensity of forestry is greater than x^*. Thus forestry can be 'bribed' to produce at the socially optimal rate. Moreover, if reindeer husbandry has property rights to the land, forestry can buy the right to use the land by paying a sum of money $\varepsilon' - \frac{dB^R}{dx}$ such that $0 < \varepsilon' - \frac{dB^F}{dx} < \frac{dB^F}{dx}$, as long as x is smaller than x^*. Hence property rights do not have any bearing on whether or not the optimal scale of production is attained. However, they do have a bearing on income distribution. These facts are made clear by Coase's (1960) theory, referred to as the Coase theorem:

Coase's theorem

If costless negotiation is possible, rights are well specified and redistribution does not affect marginal values, then:

1. The allocation of resources will be identical, whatever the allocation of legal rights.
2. The allocation will be efficient, eliminating the problem of externality.

The theorem assumes that transaction costs are zero and that property rights are well defined. Property rights do not necessarily mean the ownership of the land. It could, for example, be sufficient for the Laplanders to own a well-defined right to use the land for reindeer husbandry, while the forest industry had the right via, say, a lease contract to conduct forestry. The land could be owned by the state. Much of the conflict has been concerned with the property rights of the Laplanders which might have contributed (together with positive transaction costs) to the fact that most people perceive the conflict as being unresolved.

Another important complication in the forestry-reindeer example is the implicit assumption that there are only two parties involved. Assume that negotiations are conducted between one landowner and many reindeer owners. Now there is an incentive for a reindeer owner to conceal his true willingness to pay in the hope that others are harmed by forestry to such an extent that the 'contract' goes through. In other words, it pays for the individual to be a *free rider*. In economic theory this is often referred to as the 'free-rider problem'. If the number of reindeer owners increases, it is conceivable that the free-rider problem becomes increasingly serious, and the actual scale of forestry will approach the private optimum x^F.

Marketable Emission Permits

The idea behind Coase's theorem is that a small number of agents are able to exhaust all mutually advantageous trades, even if the institutional arrangements, apart from the well-defined property rights necessary for a market solution, are absent. However, since markets have socially desirable consequences, it may in certain circumstances be a good idea to create artificially markets which are not likely to materialize on their own. Say, for example, that we know the level of the socially desirable sulphur dioxide emissions in the Amsterdam area. How can we efficiently allocate the permissible level of emissions to firms emitting sulphur dioxide? Since the number of victims is large, and since property rights are not easy to define, the Coase theorem is not applicable. Pigouvian tax is possible, but an interesting alternative is to create a market for emission permits.

The profitable firms with high abatement costs have a high willingness to pay for sulphur dioxide emission permits, and rightly so since society's opportunity costs of getting rid of these emissions, the (marginal) profits, are high. Relatively unprofitable firms and firms with low cleaning costs have, for opposite reasons, a low willingness to pay for the right to emit sulphur dioxide, because the opportunity costs of getting rid of their emissions are low. Would not a free market for emission permits in these circumstances allocate emissions in a socially desirable manner?

To be more precise, assume that there are n firms that emit sulphur dioxide, and that the ith firm emits E_i tons of sulphur dioxide at full capacity utilization. To simplify the formal analysis, we assume that each firm's average variable cost. AVC_i, is constant up to full capacity utilization, and that the cleaning cost per-unit of emissions is constant and equal to r_i. Moreover, assume that each unit of output corresponds to one unit of emissions. Firm i's cost and revenue structure is given in Figure 2.5.

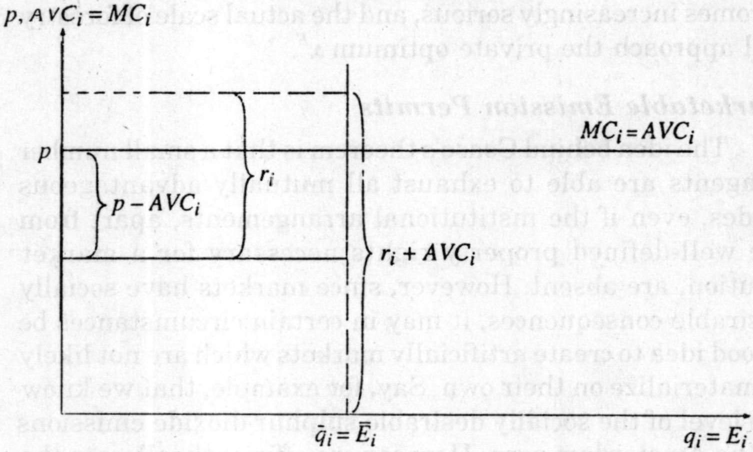

Fig. 2.5 : The cost, revenue and emission conditions in firm i.

The firm faces a price per unit of output equal to p, and makes a unit profit equal to $p - AVC_i$. In the absence of efficient charges, the firm will produce at full capacity utilization q_i and emit $E = q_i$ units of sulphur dioxide.

The firm's maximum willingness to pay for a permit to emit a unit of sulphur dioxide. SO_i, is $p - AVC_i$. If it pays this amount it will break even. If the abatement cost/unit is less than $p - AVC_i$, the best alternative for the firm and for society is to clean each unit produced (we assume that the externality is completely eliminated, that is, cleaning is perfect). However, in Figure 2.5 $r_i > p - AVC_i$ and the firm would have to go out of business if it was forced to clean its production. In other words, the firm's willingness to pay for a permit for emitting one unit of sulphur dioxide, P_{Ei} equals:

$$P_{Ei} = Min\ [p\text{-}AVC_i, r_i]\ i = 1, 2..., n. \qquad ....(6)$$

Clearly, if $r_i > p - AVC_i$ it will only be willing to pay $p - AVC_i$, since cleaning costs would in these circumstances undo all profits. If the opposite holds ($p - AVC_i > r_i$), it will not buy an emission permit if it costs more than cleaning (more than r_i), even if it can afford it in the sense that the profit per unit remains positive.

In Figure 2.6 we have ranked the firms' (marginal) willingness to pay for emissions, which also equals society's opportunity cost (the net value of production lost) to get rid of the sulphur dioxide, from the highest to the lowest.

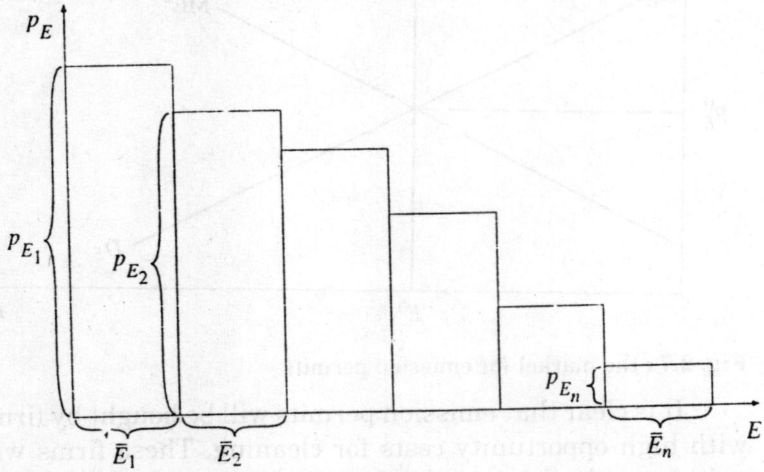

Fig. 2.6 : The willingness to pay for emissions (society's opportunity cost).

Firm 1 is willing to pay P_{E_1} dollars /ton to emit $\bar{E}_1$ tons of sulphur dioxide, while the lowest willingness to pay is P_{E_n} for $\bar{E}_n$ units. Clearly, since the diagram also represents society's opportunity cost for getting rid of the emissions, we would like to minimize this cost. Hence we would want the firm with the lowest willingness to pay (the lowest opportunity cost) to abate its emissions, while those with the highest opportunity costs sould continue to emit. This is exactly what the establishing of a market for tradeable permits would achieve. The step function in Figure 2.6 is the aggregate demand curve for emission permits. If the socially desirable level of emissions is E^0, and society chooses to sell emission permits to the same extent, there will be an equilibrium price, $P^0{}_E$ where the vertical supply curve crosses the demand curve for emission permits (see Figure 2.7), where the step function has been smoothed out.

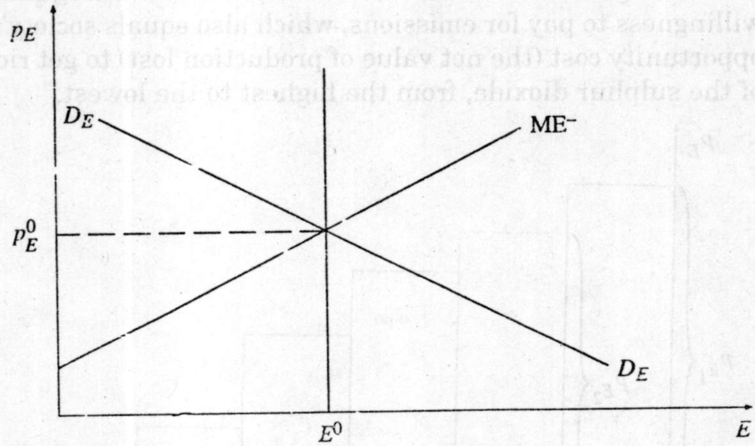

Fig. 2.7 : the market for emission permits.

It is clear that emission permits will be bought by firms with high opportunity costs for cleaning. These firms will neither cut back on production nor abate emissions. Firms which have a willingness to pay which is lower than the market price will shut down if $p - AVC_i < r_i$, or if clean output $p - AVC_i > r_i$. Mixed solutions, where the firm both abates emissions and buys emission permits, are possible in a more realistic setting where the willingness to pay for emissions depends on the volume of emissions and the cleaning cost depends on the degree of abatement.

The socially desirable volume of emission permits, E^0, is determined by the equality between the demand curve for emission permits (the opportunity cost to eliminate the permission), and the curve which measures society's marginal willingness to pay for (marginal benefit of) getting rid of the emissions (the ME – curve in Figure 2.7). The reader should note that a low emission level comes with a high marginal abatement cost (opportunity cost to eliminate the emission).

As in the tax case, it requires a great deal of information about valuations and technology in society to determine this level correctly. The best we can hope for is probably a cost-effective solution, which means that we obtain a given level (not necessarily the socially desirable level) of abatement at minimum cost. Marketable emission permits also provide incentives to improve abatement technology, since the cleaning of emissions results in compensation in terms of revenues from the sales of permits in excess of the new lower emission level.

That permits are marketable is a special advantage from a dynamic perspective. The establishment of new firms, shutdowns, and technological progress all change the demand for emission permits, and trade in permits is a flexible way of adapting to new conditions. In a growing economy, it is conceivable that Pigouvian taxes will have to be frequently adjusted. When there is a market for permits the price of an emission permit will rise automatically, reflecting the fact that the right to pollute has become a scarcer resource.

In an inflationary economy, a permit is also more flexible than a tax. Without frequent adjustments of the Pigouvian tax, environmental quality will be eroded under a tax regime. The price of a permit will simply follow the general price level, and will therefore be preferable to a Pigouvian tax.

Compared with direct controls at the firm level, a market for emission permits has the obvious advantage of being a regulation which is cost-effective.

Special Problems

What we have done so far is to design environmental policy under certainty and under the implicit assumption that the sources of pollution stem from firms active in competitive markets. The latter means that both firms and buyers are many, and that no cartels are formed; either of firms or of buyers. The former means that we are neglecting all the uncertainties of real life. However, from an environmental point of view the most important uncertainties concern the positions of the demand curve for permits (or more generally the marginal abatement cost curve) and the marginal willingness to pay for cleaning the emissions. Both problems are important in practice, and we will explain them briefly. We will also deal with externalities in a monopoly (that is, in an industry consisting of one firm), and finally with an important technicality for policy called a non-convexity, which can severely hamper the possibility of reaching a socially desirable outcome as a market solution.

Uncertainty

We will start with a case where the regulator is uncertain about society's MVP curve, which is clearer than the case where the demand curve for emission permits is uncertain. Consider Figure 2.8 which is essentially the same as Figure 2.7.

In Figure 2.8, we have highlighted the fact that one can reach the same allocation under a Pigouvian tax and a market for emission permits. The tax will coincide with the equilibrium price for the permit $t^0 = p^0_E$. Both the tax and the number of emission permits are based on the assumption that the relevant opportunity cost to clean is represented by the curve $D_E\, D_{E'}$ and that the marginal benefits from emissions are correctly represented by the ME^- curve. The policy means are, under these conditions, equivalent. Assume, however, that we have correct information on the demand curve (the opportunity cost to clean), but that the true benefit curve is ME^-_1, which means that we are under-estimating the benefits of a cleaner environment. The number of permits should be E_1 instead of E_0 and the Pigouvian tax should be t^1 instead of t^0. The social loss from excessive emissions is, however, the same under both policy regimes. It corresponds to the shaded area L in Figure 2.8 and consists of the difference between the marginal willingness to pay for emissions, and the opportunity cost to clean the emission for the excessive number of emissions $E_0 - E_1$.

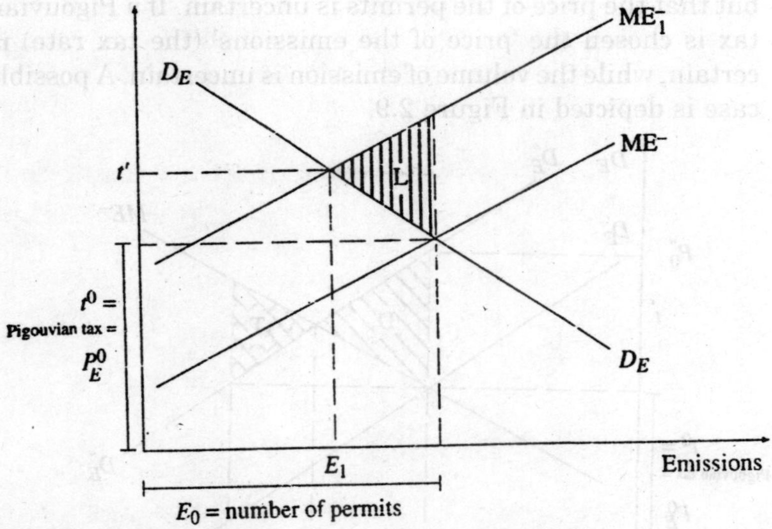

Fig. 2.8 : Pigouvian tax versus a market for permits when the ME⁻ curve is uncertain.

The equivalence between a tax and a market for emission permits, in a situation where there is uncertainty about the marginal willingness to pay, is also demonstrated for a case when the marginal benefits from reducing emissions are overestimated. The exercise is left to the reader. The intuition underlying the result, that uncertainty about the marginal benefit curve does not matter in a comparison between a tax or marketable permits, is that the policies are designed under perfect knowledge about the conditions of the acting parties in the market, that is, the firms which pay the tax or buy the emission permits.

Hence we know exactly how the firms will react to the alternative policy measures, which depend on the same beliefs (right or wrong) about the marginal benefit curve.

A more complicated situation arises when the marginal benefit function is known with certainty, but the opportunity cost to clean is not known and wrongly estimated. This means that the policy-makers, in the case of marketable permits, are able to control volume (the number of permits) but that the price of the permits is uncertain. If a Pigouvian tax is chosen the 'price of the emissions' (the tax rate) is certain, while the volume of emission is uncertain. A possible case is depicted in Figure 2.9.

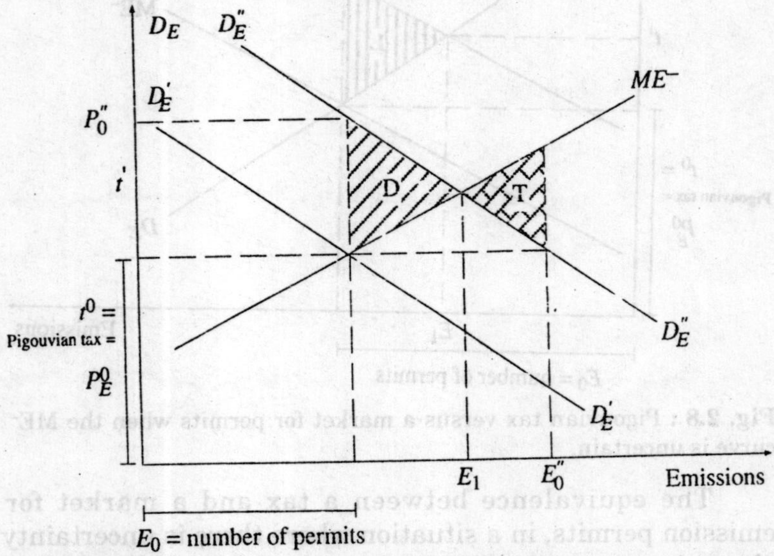

Fig. 2.9 : Comparison between tax and a market for emission permits when the opportunity cost of cleaning (the demand curve) is uncertain.

Here t^0 represents the Pigouvian tax which is based on the presumption that the marginal benefit function is ME^- and the demand function for permits is $D'_E D'_E$ E_0 is the number of permits issued under the same assumption. If the true opportunity cost of cleaning is $D'_E D'_E$, the market price for the original volume of permits becomes P''_0, and too much cleaning is carried out since the opportunity cost of cleaning is underestimated. The social loss from the mistake is the difference between the opportunity cost of cleaning and the marginal benefit function for the extra emissions needed to reach the socially desirable number, E_1. The loss is measured by the area labelled D.

Under a Pigouvian tax equal to t^0 there will be too little cleaning and emissions will be E''_0. The reason is that the Pigouvian tax underestimates the 'equilibrium price' for emissions, owing to an underestimation of the demand curve for permits. The social loss is the area T, which is the difference between the marginal benefits and the opportunity cost of cleaning for the units emitted in excess of the socially desirable level of emissions.

In this case the area T is less than D, and a Pigouvian tax is preferable to a market for permits. It is easy to construct a case where the opposite is true. What would; for example, happen if the marginal benefit curve was vertical? Clearly, since the desirable number of permits would be independent of the demand function, a market for permits would give the right answer whereas a tax would lead to excessive emissions. The opposite would be true if the marginal benefit curve was horizontal, that is, the willingness to pay for a reduction of emissions is independent of the volume of emissions. Now the tax would work perfectly, while the market for permits would allow too few emissions.

The essence is that the choice between marketable permits or taxes, from the point of view of our ignorance about the position of the relevant demand and cost functions, is an empirical matter. What can be said is that if the slope of the marginal benefit function were steep-this would be the case when one is dealing with hazardous waste-close control of quantity becomes important, and marketable permits are preferable to Pigouvian taxes. On the other hand, if the marginal willingness to pay for cleaning up the volume of emissions is constant, a tax is a better method than controlling emissions through a fixed amount of marketable permits.[5]

Imperfect Competition and Externalities

We will here deal with a special case of imperfect competition, which is known as monopoly, where a single seller faces many buyers. This means that the monopolist, when he decides how much to produce to maximize profits, takes into account what his production does to the market price. In perfect markets the firms are so numerous that they take the market price as parametrically given. Hence if a competitive solution is compared with a monopoly solution for the same market, we will find that the monopoly produces too little output and charges too high a price. If, in addition, it pollutes, this means that the social costs the pollution causes have to be internalized into the cost structure of the firm. The latter can, for instance, be done-through a Pigouvian tax. Without going into detail, we can say that the monopoly solution to externalities contains two market failures: too little is produced and too much is emitted per unit of output. The policy problem can now be illustrated by the rather sterile Figure 2.10.

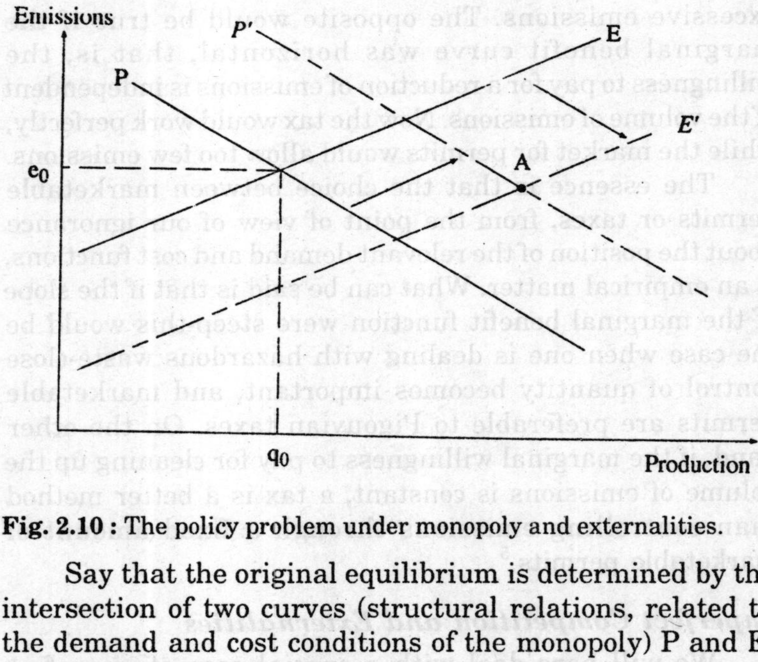

Fig. 2.10 : The policy problem under monopoly and externalities.

Say that the original equilibrium is determined by the intersection of two curves (structural relations, related to the demand and cost conditions of the monopoly) P and E. The uncontrolled solution is point e_0, q_0 and the social optimum is given by point A with emissions e_A and production q_A. To move the economy to the social optimum it is not sufficient to move only structural relation. We will have to shift both curves in Figure 2.10. Say that the E curve is shifted downwards to *E*' by a Pigouvian tax, and the P curve is shifted upwards to *P*' by a subsidy per unit of output which increases production. In this way we can make both curves intersect at point A; the socially desirable outcome.

The lesson is that, since externalities in a monopoly comprise two market failures, we will, in general, need at least two means to move the economy to the socially desirable outcome. There are, however, obvious practical problems connected with the subsidization of a monopoly, which is already making an extra (monopoly) profit from restraining production. A subsidy would mean that the profit is even larger. Therefore, in practice, a policy-marker would have to go for a second-best solution and choose a Pigouvian tax in such a way that the social surplus is maximized, given the restriction that no other policy means are allowed. In terms of Figure 2.10, he would choose to move the E curve in such a way that the crossing with the P curve is as close as possible to the socially desirable outcome. The resulting tax would depend on the demand and cost functions in a rather complex way. However, since a monopolist is producing functions in a rather complex way competitive circumstances, we will not be surprised if the second-best Pigouvian tax on emissions is lower than the corresponding first best tax. The lower output means, in itself, less environmental damage.

Non-Convexities

So far we have essentially dealt with the externality problem at the level of the firm. However, to determine the Pigouvian tax structure in an economy with externalities we must, in practice, solve a complicated control problem, where the preferences of all consumers and the technologies of all firms are involved. The idea is to pick one socially desirable outcome to aim at. The next problem is how to get there. If we use Pigouvian taxes it must be true that the optimizing actions of firms and consumers within a market economy with Pigouvian taxes lead us to a market solution which corresponds to the command optimum we aimed at in the first place. This is certainly not self-evident, and it turns out that the existence of externalities, if they are serious enough, leads to a complication which economists (and mathematicians) call non-convexities.

Consider Figure 2.11, where the shaded area represents the production possibility set of an economy where two goods are produced as outputs. A production possibility set is, broadly speaking, the feasible production possibilities of goods, given existing resources of labour, capital and energy.

Any point within the set is feasible, but the efficient points (given that more is always better than less) consist of the boundary of the set and are all, potentially, socially desirable outcomes. The boundary is called the transformation curve. Set A has the property that if we choose any two points which belong to the set, all points on a straight line between the two points also belong to the set. Such a set is called a convex set.

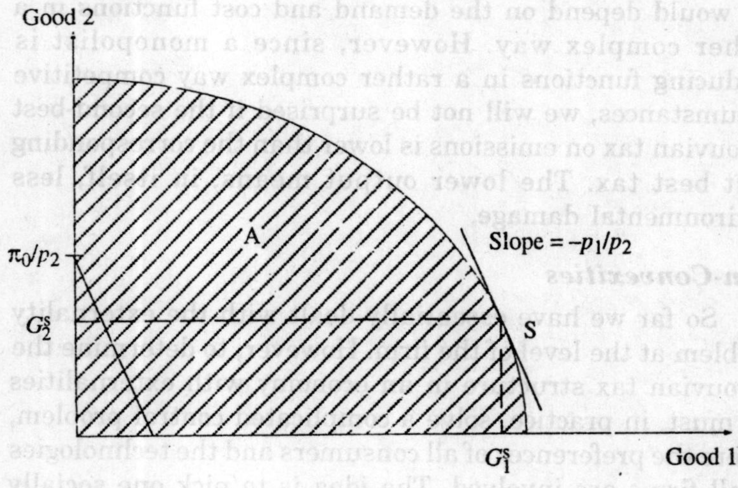

Fig. 2.11 : The convex production possibility set, A.

Say that we would like the firms to produce at point S on the boundary, where much of good 1 and little of good 2 are produced. How can we achieve this? Since firms are maximizing profits, we should make it relatively more profitable to produce good 1 than good 2, and this can be done by imposing Pigouvian taxes to make the price the producer receives of good 1 higher than the price of good 2. Given that we can neglect costs, the total profit in the economy can be written as:

$$\pi = P_1G_1 + P_2G_2 \quad ....(7)$$

where P_1 and P_2 are the prices the producer receives of the two goods, and G_1 and G_2 are the corresponding volumes. For a given level of profit, say π_0, equation 7 represents a straight line:

$$G_2 = \frac{P_1}{P_2}G_1 + \frac{\pi_0}{P_2}. \quad ....(8)$$

The slope of the line is $-\frac{P_1}{P_2}$, and when $G_1 = 0$, we have $G_2 = \frac{\pi_0}{P_2}$. The line is depicted inside the production possibility set in Figure 2.11. If we increase the profit requirement at constant prices, the line will be moved in a north-easterly direction. Given the slope we have chosen, it will eventually be a tangent at point S on the transformation curve.

A little reflection reveals that point S, the socially desirable outcome, maximizes total profit at the chosen prices. In other words, the socially desirable outcome can be reached by profit maximization within a market economy provided the price system is 'right'- which can be achieved by economic instruments such as Pigouvian taxes. Further reflection also reveals that any point on the transformation curve can be reached by profit maximization, provided that the price ratio $\frac{P_1}{P_2}$ is appropriately chosen: this can be done by, for example, Pigouvian taxes.

Consider now Figure 2.12, where we have drawn a production possibility set which is non-convex. The test is that if we choose any two points on the transformation curve, no points on the straight line between the two chosen points will belong to set B. As an example consider a steelwork and a laundry industry. The steelwork can produce 1 unit of output with 1 unit of labour: for example, 10 units of output with 10 units of labour, 20 units of output with 20 units of labour, and so on. Similarly, the laundry industry can produce 10 units of output with 1 unit of labour, for example, 100 units of output with 10 units of labour, 200 units of output with 20 units of labour and so on, if *no* steel is produced. If the steelwork does produce, the output of the laundry industry is halved. That means that, for example, the convex combination 20 units of labour, 10 units of steel and 100 units of laundry is not possible.

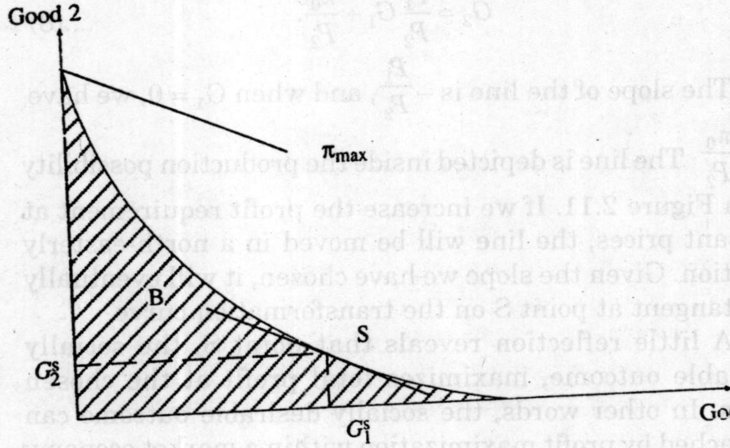

Fig. 2.12 : The non-convex production possibility set, B.

Assume that, as in the preceding figure, point S is the socially desirable outcome. To reach it through profit maximization the idea is to introduce a price system where the price the producer receives for good 1 is higher than for good 2. However, this will no longer work. Profit maximization under such a price system will move us to a corner solution in which only good 2 is produced. In other words, the socially desirable outcome cannot be reached in a market economy where competitive firms maximize profits.

The shape of the transformation curve in Figure 2.12 means that the possibility of transferring resources, from the production of good 2 to the production of good 1, deteriorates as the production of good 1 increases from zero to the corner solution.[6] This is a consequence of the fact that the transformation curve is bent towards the origin, and this is exactly what one would expect an extremely negative externality to produce, that is, when the production of one good hampers the production of another.

The relevance of non-convexities for environmental policy is that they make it more difficult to reach a socially desirable outcome. Market solutions generated from profit and utility maximization do not necessarily produce socially desirable outcomes, even if we try to correct for the negative external effects that are not handled by markets.

Valuation of Benefits and Costs of Pollution Control

To use policy instruments such as emission permits and Pigouvian taxes appropriately, one has to measure, in terms of value, the damage done by pollution (the marginal benefits from abatement) as well as the opportunity cost of control. If a change for the better is considered we want to measure the net benefits of control.

Since environmental problems are, generally speaking, caused by the lack of appropriate markets, it is natural that valuing environmental goods concerns to a large extent the valuation of goods not priced in any market. For example, how do you measure the value of the damage done by water pollution, or to put it in benefit terms: what are the benefits to be realized by controlling discharges into water from sources such as pulp and paper mills?

Many environmental good have 'public goods' properties in the sense that they are consumed by all people in almost the same quantities. The problem of valuing the benefits of, say, cleaner air is technically the same problem as valuing the willingness to pay for national defence. The theory of public goods provides both theoretical and practical answers.

Economists and econometricians have also, since the end of the 1940s, worked specifically on the environmental valuation problem. An early contribution was when the famous American economist and statistician, Harald Hotelling, in a letter to the Director of the National Park Service in 1947, said that a recreational site could be indirectly valued via the travel costs of its visitors (the travel cost method).

Hotelling's idea was that the number of trips made by the nearest visitors constitutes the demand for the site at almost zero price (travel costs are almost zero) and that the number of trips made by the visitors at large distances constitutes the demand close to the 'choke-off price' (the price at which demand is zero). In this manner, by also considering the travel costs of visitors at intermediate distances, one is able to construct a demand curve (or an MVP curve) for recreation. The area below such a curve measures the total aggregate willingness to pay for the recreational site.

The damage done by air pollution in a certain area can be indirectly valued by comparing the cost of housing in the polluted area with price of the same kind of housing in a non-polluted area. The difference between the house prices measures, in ideal circumstances (when the goods are really the same; statistical techniques can be used to control for differences), the damage done by air pollution or the value of cleaner air. This method is usually referred to as 'hedonic pricing'.

There are many environmental damages which cannot be valued by means of the travel cost method, and the hedonic pricing model cannot be used to evaluate peoples' willingness to pay for the preservation of rare flowers or animals. By asking people hypothetical questions about their willingness to pay for such things as the preservation of species, cleaner air or a cleaner river, one can hope to reveal part of the true value of these environmental goods. This method is usually referred to as the 'contingent valuation method'. It has recently become very popular, but is also much debated.

Sometimes it is wise to solve part of the valuation problem by creating new markets or by a more precise specification of property rights. By establishing a market for permits, firms' marginal willingness to pay for emissions is elicited. Unfortunately, without knowing the marginal benefits of getting rid of the emissions (the ME curve in Figure 2.7), we cannot determine the optimal number of permits. The property right idea introduced by Coase can, if one picks an appropriate situation to apply it to, solve the valuation problems at both sides of the market. In this manner, both the valuation and the environmental problems are solved simultaneously.

It goes without saying that the valuation of environmental goods is difficult, and that all the methods above have-in spite of considerable scientific progress-their disadvantages.

Concluding Comments

We have made frequent use of the terminology 'a socially desirable outcome'. We have defined it as a situation where goods are allocated in such a manner that one cannot improve the situation of one individual without worsening it for someone else. It is obvious that a socially desirable outcome has attractive efficiency properties. If it were possible to improve the situation of one individual without worsening it for someone else, it would be wrong to say that the initial allocation was efficient.

Economists refer to socially desirable outcomes as Pareto-optima, named after the Italian economist and sociologist, Wilfred Pareto, who was the first to introduce the concept in economics. Technically it is a partial ordering, since we cannot, without using subjective values, rank two efficient situations. Assume that we have a 'two-person economy' with a given endowment of goods (no production) to be allocated to the two individuals. It would be efficient to give the whole bundle to either person (there are also other efficient allocations), but we cannot say, without introducing a value judgement, to whom we should give the goods. Essentially this comes down to saying that efficiency considerations are easier for economists to deal with than distributional issues.

The nice thing about a perfect market economy (an economy with many consumers and firms which compete with one another and which contains no externalities) is that it generates a socially desirable outcome or a Pareto-optimum. If externalities are present, we have to correct the incentives in the economy in order to produce a Pareto-optimum. The ways in which this can be accomplished have been the subject of this chapter.

It has also been implicit in the discussions in this chapter that externalities are a problem, since markets which can handle all the unwanted consequences of economic interaction do not exist. The reasons for the lack of markets can be many, but when we are faced with a public goods externality such as air pollution, it will be obvious that a spontaneous market to deal with the problem is very unlikely to materialize.

On the other hand, we would expect the seriousness of externality problem *ceteris paribus* to decrease with the number of markets in the economy. This means that we would expect the environmental problems in a market economy to be less serious than in a command (planned) economy.

3

Public Policies and External Factors

Introduction

The objective of this chapter is to give first insights into the positive behavior of a government from an economic point of view. First, what environmental problems are regulated by a government? Why is it that certain problems remain to a large extent in the private sphere while other externality problems are regulated by the government? Secondly, once a government becomes actively involved, what kind of decision-making can be expected? What does democratic decision-making imply in regulating externalities? Lastly, how does the level of decision-making in the presence of multi-layer governments (federal, state or local) affect the outcome, and is there an appropriate level of government to deal with all types of externality?

Before answering the first question, we need to describe non-government solutions to externalities and their properties. Theory describes externalities as being cooperation problems, the outcome of which will depend on different parameters of the problem, such as group size, homogeneity, repetitive character, punishment possibilities and so on. These characteristics form the basis of government interventions and of private institutions such as clubs and Coasian solutions.

For an analysis of the decision-making process of the government we need to have a theory of government behaviour. There are different views, and the ones we will explore here lie between two extremes. They are somewhat complementary as they emphasize, respectively, the role of the legislative branch and the role of the executive branch of government. In the first view, the government is only an abstract decision mechanism where the private preferences of the citizens are the inputs and the collective decisions the outcome. Government decisions result from an aggregation mechanism where private preferences are summed or averaged. This mechanism is usually of the democratic type where a significant number of individual preferences are taken into account. There is a large variety of democratic institutions. We will focus on direct democracy and representative democracy with a two-part system. The properties of these decision-making mechanisms will be examined. In the second view, the government is no abstract averaging system but a group of government agents who execute, in theory, political decisions, but are, in practice, able to manipulate the decision-making completely to their own advantage. Another problem is that in most settings we do not have just one governmental authority but several government levels. In what way does the government decision depend on the level where it is taken, and is there an ideal role for federal, state and local institutions?

The organization of this chapter is as follows. Section 2 deals with solutions to externalities where there is no government intervention. Section 3 discusses the properties of different government decision rules, which are used in section 4 to analyse the political economy of instrument choice. Section 5 deals with the theory of the proper level of environmental decision-making, and in the final section conclusions are drawn.

Solutions to Externalities in the Absence of Government Intervention

In order to understand the expected involvement and role of governments in externality problems, the alternative solution without government involvement has to be explored first. It is only when a sufficient number of citizens are unhappy about the private solutions that there will be a call upon government to intervene and regulate externalities.

The classic example of a negative externality is the steel factory which emits black smoke. One of the effects of the black smoke is that it increases the operation costs of a nearby laundry. In this example we assume that the government does not intervene. There are no regulations as to the emissions of smoke nor can the laundry, as harmed party, claim any corrective action via the legal system. We will use this example to illustrate the importance of several problem characteristics such as the number of polluters and victims, the homogeneity of both types of agent and the reciprocity or sanction possibilities.

Unidirectional Externality and Only Two Parties

Assume that only two parties are directly involved in the conflict: a single steel manufacture and a single laundry-owner. Both locations are fixed. The steel manufacturer can reduce the level of smoke emissions by installing abatement equipment; the laundry can reduce the damage of the black smoke by installing protective devices. This equipment will increase the cost levels of both producers and reduce their profits. The possible outcomes are summarized in Table 3.1, where in each of the four cells the figure on the left stands for the profit level of the steel factory and the figure on the right for the level of profit of the laundry. We call these profits the pay-offs or outcomes of the game.

Table 3.1: Pay-offs in a simple pollution problem with one polluter and one victim.

	Pay-offs	*Laundry*	
		No protection	*Protection*
Steel	No abatement	20/10	20/13
Factory	Abatement	17/15	17/17

The steel factory has lower profits when it installs abatement equipment and will rationally not do so. Furthermore we assume that its operations are in no way influenced by the installation of protective equipment in the laundry. It is different for the victim (that is, the laundry). It has, in all cases, an incentive to install protective equipment because this lowers its total production costs. Hence it increases its profits by 3 units if there is no smoke abatement and by 2 if there is some smoke abatement equipment installed in the steel factory. The laundry of course prefers the steel factory to install abatement equipment because this will, in any case, increase the profit level of the laundry. The outcome one would expect, where there is no cooperation whatsoever, is that the steel factory does not install any abatement equipment and that the laundry installs protective devices. This is a *non cooperative equilibrium* for both parties: whatever the laundry does, the steel factory has no incentive to install abatement equipment. On the other hand, the laundry has an incentive to install protective equipment in any case. This is not a Pareto-optimum.[2] One could get lower overall production costs and higher total profits by having the steel factory abate smoke emissions. Total profits for the steel manufacturer and the laundry would then be 17 + 17 = 34 rather than 20 + 13 = 33 in the non-cooperative equilibrium. This optimum situation can be reached if both parties involved make binding commitments. Then we reach a *cooperative equilibrium* where the total benefits for all parties are maximized and where the extra benefits from the non-cooperative equilibrium are used to bribe all parties so that they will participate. Here it is in the interest of the laundry to pay the steel factory an amount of maximum 4 units (= 17 – 13) in order to make it abate smoke emissions. The steel factory has an interest to comply with this request if it receives at least an amount equal to 3 units (= 20 –17) for the abatement. This is the Coasian solution.

Even though a Pareto-optimum is reached, the laundry will be interested in having the *government involved*. The laundry can improve its welfare position if the government can enforce the right of the laundry to a clean environment. In this way the laundry can have the steel factory pay for its own abatement investments so that the laundry and the steel factory both receive a profit of 17. It is important to realize that the call on government intervention may result from efficiency considerations (enforce a cooperative type of solution) as well as from redistribution motives (making the other pay for a benefit I will enjoy). It will, in general, be impossible to separate both aspects.

Externality with Many Victims and One Polluter

Let us change the previous example by adding nine laundries but maintaining just one steel factory. Table 3.2 summarizes the profit opportunities for both parties. This table has been constructed assuming that all laundries are identical and no economies of scale[4] are in place. Profits per laundry are then equal to one-tenth of the profits in the Table 3.1

Table 3.2: Pay-offs in a pollution problem with one polluter and many victims.

	Pay-offs	*Each laundry (ten in total)*	
		No protection	*Protection*
Steel	No abatement	20/1.0	20/1.3
Factory	Abatement	17/1.5	17/1.7

As before, the steel factory has no incentive to abate emissions and it is in the interest of each laundry to invest in protection equipment. This is the non-cooperative equilibrium. The Pareto-optimum and cooperative equilibrium will again be that the steel factory abates smoke emissions. The highest total profit to be obtained is 17 + (1.7 × 10). However, in contrast with the 'one laundry' case, this cooperation will be hard to realize because each of the laundries has to agree to pay for the abatement equipment of the steel factory. It is in the interest of each laundry not to contribute and to have the others pay for the abatement equipment because it will benefit from the smoke abatement anyway. This is called *free riding*.

A cooperative outcome is unlikely but not impossible. The probability of a cooperative outcome increases when the situation is *repeated* regularly so that each laundry can make its contribution to the abatement infrastructure dependent on the contributions of the other laundries in the previous rounds. If each laundry makes exactly the same contribution as the other laundries in the previous round ('tit for tat' strategy) and if each laundry starts with the correct contributions (at least (20 – 17)/10 = 0.3), then an efficient outcome will be reached. A factor which increases the likelihood of a cooperative outcome is the *homogeneity* of the victims. If all laundries are of a different size and experience different degrees of damage, it is difficult to divide the cost of the abatement equipment correctly among the laundries, which makes cooperation less likely.

All the laundries are interested in having the government involved to secure their property rights to a clean environment. Since a cooperative (and Pareto-efficient) solution is highly unlikely in the large number case, government involvement is important for efficiency reasons. The government can force a cooperative outcome and increase total profits.

For completeness we can mention the problem where there are many polluters and only one victim. In this case the victim will have an incentive to pay for the pollution abatement equipment in the different steel factories. This situation is not equal to the case with many victims since there is no free-rider problem.

Externality with Many Victims and Many Polluters

We now change the previous example by adding nine steel factories. Table 3.3 now summarizes the profit opportunities for both parties. It is assumed that the laundries as well as the steel factories are identical and that no economies of scale are in place.

Table 3.3: Pay-offs in a pollution problem with several polluters and several.

	Pay-offs	*Each laundry (ten in total)*	
		No protection	*Protection*
	No emissions abatement by any factory	2.0/1.0	2.0/1.3
Steel factory	.....	.....	.....
	Emissions abatement by ten factories	1.7/1.5	1.7/1.7

The different steel factories may have different strategies. Each of them may either abate or may not abate emissions. The number of abaters will determine overall smoke emissions and consequently the damage for the laundries.

The non-cooperative equilibrium will be, as before, protection at the laundry level and no abatement by the steel factories. The cooperative equilibrium will be abatement by all steel factories and protection in all laundries.

As in the previous example, a cooperative outcome is unlikely but not impossible. Repetition of the game and homogeneity will increase the likelihood of cooperation. Because there are several polluters a cooperative outcome is more difficult to obtain if the group of polluters is not homogeneous and if monitoring of abatement efforts is costly.

We have discussed at length the traditional example of negative externalities. This example can be extended to other problem areas and to other types of agent, such as households, firms, local governments in a federation and so on. We can also mention global warming; a problem to which all nations in the world contribute and whereby all people can be considered polluters (emitting the greenhouse gas carbon dioxide). There is no world authority enforcing an agreement, so that the outcome of an agreement depends on the balance of the required efforts, on the monitoring and on the attitudes of the different nations. The pollution of a river by different firms is another example where, depending on the number of firms involved and their homogeneity, cooperation is likely or unlikely. If these firms have the same types of activity and discharge, efforts to reduce pollution can be agreed upon and monitored more easily.

Conclusions

If there is no government intervention, some externality problems are solved by private cooperation. This can take the form of an implicit or explicit contract. Cooperative solutions will be more likely if the number of victims is small, if the group is homogeneous and if the externality situation is repeated. If cooperation is absent the outcome will, in general, be Pareto-inefficient.

Government intervention can be invoked to correct the private outcome in two ways. It can, even if there is an efficient cooperative solution, be forced to correct the distribution of the property rights to the benefit of one of the parties. This has more to do with the distribution of resources than with the efficient use of resources. Secondly, it can be asked to establish a mechanism to enforce cooperation between the parties concerned, which means that a higher level of pollution abatement is enforced so that all damages are taken into account. In the next section we analyse how governments actually take decisions.

Government Decision Rules

Views on the Government

There are different views on the government and these will determine the interpretation of the outcome of public decision-making. We will start with a theoretical and idealized view of the government and end with an extreme Leviathanian view. The whole spectrum of views is necessary to understand present public decision-making. We will first concentrate on the general properties of the public decision mechanisms. The political economy of environmental policy-making will be discussed later.

In the idealised view we start with defining normative criteria to judge alternative outcomes of a public decision-making process. Next we will analyse two theoretical mechanisms to solve cooperation problems: the unanimity rule and the preference revelation mechanism. In the more realistic views of public decision-making, the state is considered to function as simple democratic decision-making processes. Government decisions are then a mere translation of many individual preferences into one collective decision via direct democracy or representative democracy. Finally, we introduce complexities such as an executive branch of government, interest groups and imperfect information, which will lead us to the other end of the spectrum: the Leviathanian view of government.

Idealized Public Decision Rules

Normative rules : In order to judge the functioning of different rules we need welfare criteria. We need criteria to determine that one outcome (in terms of income, pollution, income distribution and so on) of a particular decision is better than the outcome of another. Economists usually employ consequentionalist criteria in which only the final outcome and not the process leading to the outcome matters. Moreover, economists usually rely on the appreciation of the outcomes by each of the individuals concerned. To judge alternative outcomes, individuals use criteria such as *utility functions*. This is a function, constructed by economists on the basis of observed choices of an individual, that always attributes a higher (utility) value to an alternative that is preferred more. Utility values have no absolute meaning, only the relative value counts for an individual. To compare a reference outcome $\left(x_1^0,..........,x_N^0\right)$ in which the different values for x represent the utility of the N individuals involved with an alternative outcome $\left(x_1^1,..........,x_N^1\right)$, economists use either the Pareto rule or a social welfare function (SWF).

The *Pareto rule* is a partial ordering of outcomes: only if outcome 1 is better (that is, has a higher utility value) or worse than outcome 0 for all individuals can we say that 1 is better (worse) than 0. This is a partial ordering because all pairs of alternatives for which this property does not hold can simply not be compared. In reality there will be many such pairs because it is sufficient to have only one individual suffering a small loss to have this non-comparability. The SWF defines a complete ordering of alternative outcomes. Because in an SWF weights are attached to the utility of the different individuals, one is able to rank also alternatives where some individuals gain and others lose. One of the properties one can build into an SWF is that it respects the Pareto rule. Assume that only one individual is affected by two alternatives A and B, and that this individual prefers situation A to situation B (the utility of A is higher than that of B). The Pareto rule then requires that the SWF also attaches a higher value to situation A than to situation B, so that situation A will indeed be chosen. This will be the case when the utility of each of the different individuals receives a positive weight in the SWF.

The SWF and the Pareto rule are normative concepts here: how should one compare different alternatives? Each individual can have his normative views, and within a group these views will differ. In a democratic setting these views will have to be added or averaged in some way to arrive at a political decision that the government will execute. This problem of aggregation will be discussed later. Let us continue with another setting: if one accepts the Pareto rule, which ideal mechanism guarantees that with externalities the Pareto-superior alternative will always be selected? By restricting the potential solutions to the Pareto-optimal solutions (which improve the utility of all individuals), one rules out important redistributions between economic agents and concentrates the analysis on efficiency improvements starting from a given initial distribution of welfare. Indeed, the alternatives where the very rich lose some of their utility and the very poor have their utility increased cannot be evaluated.

Ideal public decision mechanisms : We return to our example with steel factories emitting black smoke which are surrounded by laundries. The outcome without government intervention is no abatement and a high level of smoke. The steel factories will incur a net loss if they abate, so there will be no abatement. This situation is inefficient and can be improved because the cost of smoke abatement for the steel factories is smaller than the benefits of smoke abatement (saved cost) for the laundries. The inefficiency persists because the laundries do not cooperate to bribe the steel factories: when it comes to collecting the individual contributions each laundry expects the others to contribute. One of the roles of the government is to help to solve such cooperation problems if there are possibilities of improving the utility of all individuals.

It is reasonable to start with an exploration of the *unanimity rule* as a potential ideal public decision-making mechanism. If there are alternatives in which all individuals gain, everybody will agree to the Pareto-superior alternative. The only role of the government is to find that alternative and to make the people accept it. A unanimity decision rule is attractive but will not work, because the government has to know the exact damage of the smoke to the different laundries in order to be able to design an alternative (here including a set of individual contributions and a level of abatement of the smoke) which is preferred by all individuals. The government does not have this information and it cannot rely on the parties involved that have to reveal this information. Strategic answers will be given: parties will try to make the government believe that they gain less from smoke abatement than they really do, in order to lower the contribution that will be required.

We will analyse majority rule in more detail later and will see that this does not solve the problem either. A series of decision-making mechanisms has been designed to solve this preference revelation problem. In these *preference revelation mechanisms* all participants have an incentive to reveal their preferences which show the true costs and benefits of abatement. To see how such a mechanism works we use an example with two steel factories and two laundries. There are only two alternatives: impose smoke abatement or do nothing. The net gains of these two alternatives are summarized in Table 3.4.

Table 3.4: The preference revelation mechanism applied to a pollution problem with two polluters and two victims.

Pay-offs	*Abatement*	*No Abatement*	*Incentive tax*
Steel factory A		20	
Steel factory B		10	0
Laundry C	28		8
Laundry D	22		2
Total	50	30	10

If no abatement is imposed by the government, the steel factories save the abatement costs (20 + 10=30). If the government imposes abatement, laundries C and D will gain, respectively, 28 and 22. The government has to choose between the two alternatives but does not know the abatement costs and benefits. The government may tackle this problem by using a preference revelation mechanism as follows: all parties are asked to reveal how much they are prepared to pay for one alternative rather than the other, which will result in the alternative with the highest 'total willingness to pay'. This mechanism in itself will not work because the laundries will exaggerate the benefits of abatement and the steel factories will exaggerate the costs of abatement. Typically, the laundries will announce a very high willingness to pay for the abatement alternative (say a few thousand dollars) and the steel factories will show a similar high willingness to pay for the 'no abatement' alternative. As long as these statements are not linked to actual payments this problem will remain. Therefore an 'incentive tax' is added which is to be paid by the different parties. The only function of this tax is to induce the parties to reveal their preferences and not to finance abatement or to redistribute any revenues. In fact, the incentive tax must not influence the answers. This is why the proceeds should be used for other projects.

The incentive tax is not a simple function of the willingness to pay of each party. First, an individual only has to pay an incentive tax if his answer is decisive. If so, he has to pay a tax equal to the net costs imposed on the other participants by his statement. First we calculate the incentive tax assuming that all participants reveal truthfully their pay-offs. Later we will show that revealing truthfully is in the interest of each participant. In Table 3.4 participant A does not have to pay any incentive tax because his answer does not alter the decision in favour of abatement: without A's revelation that his benefits are 20 in the non-abatement option, the total pay-off of the abatement option as indicated by participants C and D (50) would still be higher than the total pay-off of the non-abatement option revealed by the other participants (10). Laundry C has to pay an incentive tax of 8. This incentive tax equals 8 because without the pay-off declared by C, the non-abatement option would win with 30 to 22. This means that C's revelation is decisive in favour of the abatement option and causes a net loss for the other participants of 8, because the other participants prefer in his absence 30 to 22. Similar reasoning leads to the incentive taxes to be paid by B and D.

We can now demonstrate that with this system of incentive taxes it is in the interest of the participants to tell the truth. Assume that A would like to reverse the decision and over-reports his pay-off (or cost savings) by stating 50 rather than the actual 20. He has now the casting vote and will have to pay an incentive tax of 40 (= 50-10). But A will not reveal 50 because his real gain will be -20 (= 20-40) (the real pay-off minus the incentive tax paid) which is lower than the net pay-off (20) he can obtain by telling the truth.

The same holds for the laundries. C will not over-report because this will not change his incentive tax payments. It is only when C reports a benefit lower than 8 that he can change his incentive tax payments: his revelation is no longer decisive because the 'no abatement' alternative wins. This, however, is not in the interest of C because in the 'no abatement' alternative he loses an amount of benefits equal to 28. So the best strategy for all parties is to reveal truthfully their costs and benefits.

This incentive mechanism can solve cooperation problems. Up to now it has remained largely a theoretical construction that has not been used for important public decisions. It is probably too complex and not transparent enough to be used widely. Furthermore, it confines the externality or public goods problem to a pure efficiency or cooperation problem, while most people want the equity aspects to be taken into account in all decisions concerning allocations, including those that have to do with the environment.

Direct Democratic Decision-Making

Decision-making by majority is the most widespread public decision-making mechanism. One usually distinguishes between direct and representative democracy. In the latter system the voters elect representatives who then take the final decisions. We first concentrate on direct democracy and assume a government in its purest form: the decisions taken by the voters are executed (there are no intermediaries such as Members of Parliament and bureaucrats).

In order to examine the properties of this mechanism we use the siting of an electric power station as an example. A democracy has to decide on the site of a new power station within the country whose boundary is depicted in Figure 3.1. Three groups of citizens, all groups of equal size, live in that country in cities A, B or C. the benefits of the power plant are the same for all citizens irrespective of locality. The emissions of the power plant are, however, a discomfort for all citizens and very with the distance of the respective cities to the plant. This is illustrated in Figure 3.1.

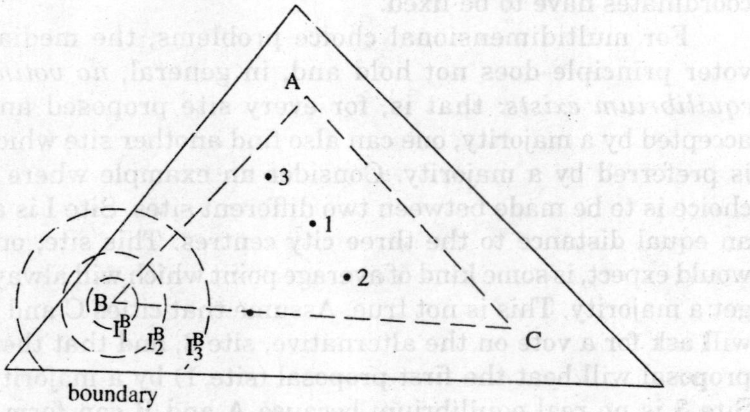

Fig. 3.1 : Cycling in majority voting systems.

The preferences (or iso-disutility lines with respect to the emissions of the power plant) take the form of concentric circles around cities A, B or C; for city B these are the level curves I^B. Iso-utility lines are lines where the utility function has the same value. So city B is indifferent to all locations on the same I^B curve. Points on I_3^B are preferred to points on I_2^B, points on I_2^B are preferred to points on I_1^B.

We want to know the properties of simple majority voting in this case. There is a widespread belief that the preferences of the pivotal voter or *'median voter'* are decisive. The idea is that the voter with the median preference determines whether there is a majority or not: there is, by definition, an equal number of voters on the left and on the right, which makes the median vote decisive. This principle holds for unidimensional problems (legislation problems along one line). However, most public decision-making problems and certainly environmental problems, are multidimensional problems, such as the localization problem. In many environmental problems we have not only the dimension, 'which level of environmental quality we want to reach' but also the dimension, 'who has to pay for it'. This corresponds to the selection of a point in the localization problem where the vertical and horizontal coordinates have to be fixed.

For multidimensional choice problems, the median voter principle does not hold and, in general, *no voting equilibrium exists:* that is, for every site proposed and accepted by a majority, one can also find another site which is preferred by a majority. Consider an example where a choice is to be made between two different sites. Site I is at an equal distance to the three city centres. This site, one would expect, is some kind of average point which will always get a majority. This is not true. Assume that cities C and B will ask for a vote on the alternative, site 3, and that their proposal will beat the first proposal (site 1) by a majority. Site 3 is no real equilibrium because A and B can form a coalition and propose site 2. This will defeat proposal 3. Proposal 2 can again be beaten by another proposal, and so on. This property, also called *cycling*, implies that the outcome of the majority rule can be influenced by *agenda-setting* (the sequence of the voting). The setting of the agenda is important because by careful agenda-setting one can find a majority for almost all sites. This means that the setting of the agenda becomes an important element in reaching a decision. A second important characteristic of a majority voting system compared with ideal decision-making rules is that it does not necessarily observe the Pareto rule. McKelvey (1976) has shown that one can construct voting paths where proposals which imply more damages for the three cities are selected by majority rule decisions.

The property that there exists no alternative preferred by a majority (that is, one particular alternative cannot be beaten by a majority) conflicts with our intuition and day to day experience of politics. We will see this problem solved when we discuss the more realistic probabilistic voting theories for a representative democratic system. Cycling and agenda-setting remain, however, important for all small group decision-making problems where decisions are taken by majority rule.

Representative Democracy

A direct democratic system is an exception in the Western world-only Swiss referenda come close to a direct democracy. Most political decision mechanisms have a two-layer structure. The individuals elect representatives on the basis of a broad programme and the elected representatives are supposed to carry out this political programme. We will restrict ourselves to discussing two-party systems. In order to judge the outcomes we need to study the behavior of the (candidate) representatives, of the voters and of the elected representatives.

The most widespread assumption on the behaviour of candidate representatives is that they try to formulate policies in order to win elections rather than to win elections in order to formulate policies. If there are only two political parties we can use the electric power plant example to illustrate the behaviour of the two parties. Assume that the first party proposes location 1 in the hope of maximizing votes with its centrist position. The second political party can then propose location 3, but each of these proposals can always be beaten by a proposal of the first party such as solution 2 and so on. Consequently, a real equilibrium does not exist and the outcome will depend on other institutional factors such as agenda-setting. Another element to assume is that each party, once in power, has to defend the progamme it has carried out at the next election. This type of democratic system leads to an historically determined sequence of equilibria where political parties alternate in office and carry out the party programme. This can lead to a jigsaw type of environmental policy: each new government proposes a different mix of environmental policies.

Up to now we have assumed two important implicit factors. We have assumed that each party, in selecting a platform, knows exactly the distance between its proposal and the preferred point of each voter. Secondly, we have assumed that each voter is perfectly informed about the proposal of the two political parties. When we relax these two assumptions we have a *probabilistic voting system*. Each political party now tries to maximize the expected number of votes it will receive. The probability of receiving the vote of an individual is highest when the ideal solution from the voter's point of view is proposed. This can be seen as the top of the probability mountain of that individual (the volume of each of these mountains equals 1). If a proposal is situated at a lower altitude of the probability mountain of that individual, the probability that the proposal of the political party will receive the vote of the individual will be smaller than for the other proposal. Each party will try to make a proposal which maximizes the cumulative height of the different probability mountains.

In this interpretation of the majority decision-making process, cycling disappears, an equilibrium exists and both parties will propose (almost) the same programme, which will correspond to the highest height of the cumulated probability mountains of the different voters. Not only will an equilibrium exist, but it will also have some desirable properties: it will always be Pareto-efficient. This means that anomalies such as, for instance, proposal A preferred to proposal B by all parties, but yet defeated by proposal B after a voting cycle (see section 3.2) will not occur. Furthermore, the equilibrium corresponds to the maximum of an SWF (for example, of the type $W=U_1+U_2+.....+U_N$). This corresponds to the idea of democracy where each vote has an equal weight in the decision-making or where the final decision is an average of individual preferences.

More Complex Decision-Making Processes

We need to make the decision-making process much more complex in order to have a more realistic representation of public decision-making processes. We need to introduce interest groups, bureaucratic influences and so on.

Interest groups (groups of voters with the same preferences) can influence the decision-making process if they can convince the political parties that they know the preferences of their members and that these preferences are homogeneous. This will give the preferences of the members of an interest group a more than proportional weight in decisions. Interest groups (Greenpeace, Friends of the Earth and so on) are important factors in environmental problems. By their actions they try to promote safeguarding the environment and, in this, defend the interests of those people with a higher willingness to pay for a clean environment. Olson (1965) has mentioned some problems with respect to interest groups. The basic problem of these associations is again a free-rider problem. As long as the association cannot confine the benefits of an improved environment to the members who support the association, there is an interest for each individual not to become a member and to leave the efforts to others. Free-riding behaviour of individuals has been the subject of experiments which have showed that in large groups not all individuals free ride, and that free riding is less important in the first rounds of the game but develops in the next rounds when there is no information about the contribution of the different individuals. Interest groups can try to solve this problem by offering other exclusive benefits to their membership such as trendy clothes or other articles.

In the views of the political process that has been used here, there was no role for the executive branch and all political decisions were carried out perfectly. When information about the feasibility, benefits and costs of certain programmes is imperfect (as is often the case in environmental issues) and/or these actions have to be performed by the administration (such as environmental licences to operate an industrial plant), the *bureaucracy* has a lot of power to influence public decision-making. The bureaucracy pursues goals other than the supply of environmental goods at the lowest cost. They may be interested in more 'power': larger budgets, offices, manpower and so on, or they may try to favour some of the solutions for which they have a subjective preference. This problem is not confined to environmental agencies; there is competition from several departments (environment, defence, welfare, infrastructure and so on) for the same budget. However, because the outputs and performances of the different departments are difficult to compare, the existence of 'bureaucratic' competition for the same budget does not guarantee a good allocation of resources.

Leviathan View of Government

If you push the power of bureaucracies sufficiently far, political competition becomes irrelevant for the ultimate decision-making. This is a Leviathanian view of the government. The idea is that the government's primary objective has become revenue maximization. The government revenues are then used to fulfil the bureaucrats' aims. Two types of brake have been proposed against this type of behaviour. The first is to build in constitutional rules that limit the functions of government, the importance of the public budget and so on. A second instrument that has been proposed is the earmarking of fiscal revenues to certain types of expenditure programme. This Leviathanian behaviour has to be seen as a very slow process where bureaucracies enlarge their power until a tax or debt revolt breaks out and the control of voters over government is momentarily reinforced.

Political Economy of Instrument Choice

The large diversity of factors affecting public decision-making mechanisms means that there is no well-developed positive theory of environmental policy-making. A positive theory should be able to explain why some issues are regulated, which instrument will be used for that regulation and what its effects will be.

To predict whether a certain environmental problem will be regulated or not we can draw on the basic economic interests explained in discussing the outcome without government involvement (see section 2). There are two main forces. When there is a large number of victims of pollution, none of the victims will voluntarily make attempts to have a sufficient level of abatement; only the government (pressed by interest groups) can bring about an efficient level of abatement. The second force pressing for government intervention, even in cases with a small number of parties involved, is the attempt of the victims to make the polluters pay for the abatement costs.

The precise outcome of a government intervention is to a large extent determined by the type of instrument used for environmental policy. Several attempts have been made to explain the much larger use of standards compared with emission fees. The first explanation starts from the idea that polluting firms in an industry are better organized than the victims (for example, consumers and different types of other firms). Consequently, they are better able to influence the government (environmental agency, Members of Parliament) than the victims. For the same level of pollution, the firms will prefer standards because these yield higher profits for two reasons. With standards firms do not have to pay taxes on the remaining emissions. Moreover, when the form of standard can be chosen by existing firms, standards may be a barrier for new firms to enter the market. A second explanation has to do with the symbolic value of standards for environmental interest groups and politicians. For the general public, standards provide a clear and apparently more certain outcome in terms of pollution reduction. Marketable permits and emission fees are perceived as negative as they give a 'right' to pollute. The symbolic value of standards can also be an explanation for the fact that standards are frequently not monitored adequately and are not met because they are too stringent in the short term. The answer can be found in the pay-off for the different interest groups. Environmentalists want stringent standards as a symbol for higher environmental quality and a sign of their efficiency. Industry will be strongly opposed to stringent standards but can live with badly enforced standards. A third explanation has to do with the interests of the environmental agency. It can expand its importance and manpower by imposing very detailed and stringent emission standards even in cases where a simple emission fee would have been much more efficient. Of course, if the proceeds of the emission fees do not go directly to the Treasury but can be used by the environmental agency, the preference for standards could be reversed to fees.

The Proper Level of Environmental Decision-Making

The solution to an externality problem can be left to individual agents, can be decided at the level of local or national government, or can be settled on a European or worldwide (United Nations) basis. All the solutions are likely to be different, because different interests are involved in the decision-making process. At lower levels we might have only polluters or only victims, while at higher levels both categories will be present. Moreover, the decision-making mechanisms used are different. They can range from referenda at local level with two-party parliamentary systems, qualified majority, or unanimity at supranational level. As can be seen in the previous sections, the outcome (equilibrium) of some of these decision-making mechanisms depends strongly on institutional details or on the trajectory of past decisions. As some of these mechanisms do not even possess an equilibrium or depend very strongly on institutional details or the trajectory of past decisions, it is impossible to compare accurately the merits of decision-making at each level. The most we can do is to discuss some general principles. One of the most influential ideas is Oates's decentralization principle (1972), discussed first. A second principle has to do with the choice of policy instrument: each level of decision-making has its own constraints on the types of instrument which can be used.

The Decentralization Principle

Oates's decentralization principle implies that each environmental problem can best be solved at the lowest possible level where benefits and costs of action go together. Important advantages of decentralization to the lowest possible level are the possibilities of tailoring the solutions to local circumstances, such as preferences and costs. The lower cost of information and control is another potential advantage. This type of solution has 'internalized' the externality problem at the lowest possible level.

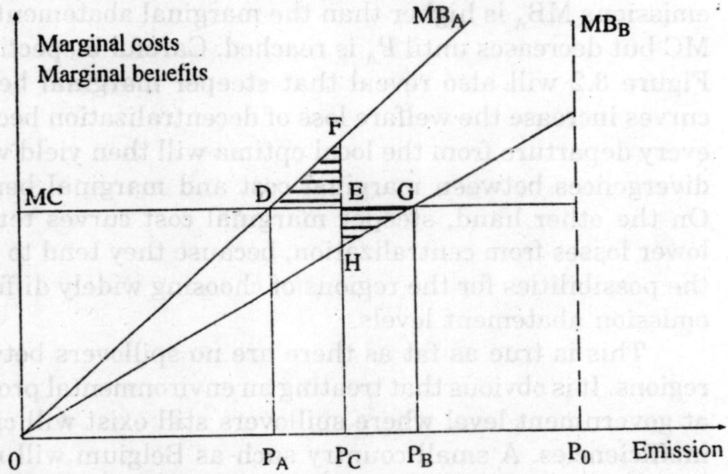

Fig. 3.2 : Advantage of decentralized decision-making.

The advantage of decentralization is illustrated in Figure 3.2. Imagine two regions (A and B) which have to decide on the level of abatement of a pollutant with only a local impact (for example, noise). The marginal abatement costs are the same and constant in both regions: the line MC. Assume that with the decentralized solution both regional governments maximize the welfare of their constituency and maximize the difference between total benefits and total costs of pollution abatement. This maximum is reached when the marginal benefit of emission reduction equals the marginal cost. Both regions have a different preference for environmental quality: region A has marginal benefit curve MB_A and region B has marginal benefit curve MB_B. The regions will select a different pollution level: P_A and P_B, respectively. We can compare this solution with a 'centralized' solution where the central government has to select the same level of pollution for both regions. This will be a level P_C somewhere between both optimal levels. The cost of centralization (or uniformity) are the areas DEF and GEH. Area DEF is a loss for region A because it corresponds to the excessive pollution inflicted on the region: at level P_C the damage of the last unit of emissions MB_A is higher than the marginal abatement cost MC but decreases until P_A is reached. Careful inspection of Figure 3.2 will also reveal that steeper marginal benefit curves increase the welfare loss of decentralization because every departure from the local optima will then yield wider divergences between marginal cost and marginal benefit. On the other hand, steeper marginal cost curves tend to lower losses from centralization, because they tend to limit the possibilities for the regions of choosing widely different emission abatement levels.

This is true as fat as there are no spillovers between regions. It is obvious that treating an environmental problem at government level where spillovers still exist will create inefficiencies. A small country such as Belgium will make insufficient efforts to abate acidifying emissions because only 10 to 20 per cent of its emissions will damage its own territory. As each environmental problem has its own zone of influence, the decentralization principle would require the creation of a public decision-making institution for each category of environmental problem. This is, of course, impossible, and an imperfect matching will have to be made between existing government structures and environmental problems. The existence of spillover effects has to be solved either by bringing the decision to a higher government level or by bilateral or multilateral agreements.

The Choice of Instrument

The choice of policy instrument to implement the desired level of pollution abatement can, in principle, also be decentralized to that government level where costs and benefits coincide. There are, however, additional complexities because the use of certain environmental instruments at a lower government level can hamper the realization of government goals at a higher level. More specifically, the choice of instrument at local level can interfere with the free trade of goods and services in a federation or can be used deliberately to distort competition in favour of local interests. These issues are dealt with in Chapter 9 on instrument choice, and more specifically in Chapter 12 which discusses environmental policy in the European Union.

Conclusions

In this chapter we have explored how public decision-making on externalities is carried out. There are basically two types of reason that citizens use for invoking government intervention for externality problems. Government intervention can be necessary to overcome the insufficient level of pollution abatement which will result when a large number of victims is incapable of raising sufficient funds to pay the polluters for their abatement actions. The second reason why government intervention is requested by citizens is the definition and enforcement of their rights to a clean environment: the victims can pay the polluters to make them abate but only public intervention can make the polluters pay for their own abatement and damage done to the victims.

There is no easy positive theory of the economic aspects of public decision-making. There are two major difficulties. The first problem is that for most voting issues no single equilibrium exists so that the actual outcome depends on the history of previous decisions and on institutional details such as agenda-setting. The second difficulty is the power of intermediaries such as politicians and bureaucrats in the decision-making process. Their interventions, made possible by their foreknowledge, do not fully reflect the preferences of the voters.

The precise outcome of a government intervention is to a large extent determined by the type of instrument used for environmental policy. The fact that standards are more largely used than emission fees can be explained by different factors. The polluting industry is, in general, better organized than the victims and can, as a pressure group, more efficiently press for the less costly standards.

Standards also have a high symbolic value for the victims of pollution and for environmental pressure groups. Lastly, environmental agencies can boost their importance and manpower by regulating via detailed emission standards rather than by emission fees.

There are different levels of government. It is commonly accepted that an environmental problem should be treated at that level where all costs and benefits can be internalized. This has the advantage that the environmental quality can be differentiated according to local preferences. The main disadvantage of decentralized solutions to pollution problems, namely the existence of transboundary pollution, is absent as long as the polluters and victims belong to the same constituency. Decentralized environmental policy-making cannot, however, make use of the full range of environmental instruments because some may hamper the realization of other objectives such as market integration.

4

Dynamics of Environmental Resources

Introduction

All of us form models of the world in our heads which help us to understand what happens around us and to imagine or predict the future consequences of choices made in the present. We begin modelling by considering what is certain and what relationships exist at that moment. It is only natural that we simplify. Dynamics and uncertainty are two aspects of modelling which favour a more realistic point of view. They are intricately related; complicated dynamics may lead to uncertainty about relations in future time periods and uncertainty about current relations is compounded in time by dynamics.

In this chapter, each section introduced general terms and approaches, continues by exploring issues more specific to the economic analysis of environmental problems, and concludes by restating the aspects relevant for decision-making. Section 2 deals with dynamic issues in environmental economics. In section 2.1 a brief introduction to dynamic analysis includes stability, attractors and other solutions, exponential increase, thresholds, bifurcations and irreversibilities. Section 2.2 covers the issue of discounting the future, a procedure which provides a current value for future states and is the subject of much debate in the field.

In section 2.3 a descriptive account and a critique of the optimal growth model which underlies much of dynamic theoretical analysis in environmental economics are presented. In section 2.4 the main implications of dynamic analysis for policy questions are summarized.

Section 3 deals with uncertainty. In section 3.1 a discussion of the limits to knowledge from various sources of uncertainty is provided, along with simple examples of modelling with random variables. In section 3.2, two case studies demonstrate techniques of dealing with uncertainty in larger models for which analytical solutions are not possible. Section 3.3 treats the control problem introduced in section 2.3 in the case of uncertainty and noise. The stochasti and adaptive control methods are briefly described. Section 3.4 concludes with policy intuitions and implications relevant to the issue of uncertainty, a final note and further reading suggestions.

Dynamics

Introduction

Dynamics is the study of the movement of variables through time. In environmental economics we are concerned with economic variables such as the stocks and flows of gross national product (GNP) and total capital (production capacity), interest rates, prices and 'utility', as well as environmental variables such as stocks of resources, flows of emissions and the state of the environment. We attempt to model the pertinent relations between variables and changes in variables through time, by formulating equations and systems of equations. In some cases these can be solved and, from the exact solution, the dynamic behaviour of the system is exactly (quantitatively) determined, but for many nonlinear system this is not possible. For unresolvable systems the qualitative behaviour of the system can be determined, and it is often the case that these are the aspects of interest. If, however, the purpose of the model is to give quantitative answers for policy analysis, and the system cannot be resolved, numerical simulations on computers can provide the necessary information. One should usually improve familiarity with the model and its dynamics by first acquiring qualitative knowledge from analysis before turning to numerical simulations.

A most important qualitative characteristic of a dynamic system is its stability. The dynamics of a system may be categorized as stable, that is, the variables remain within a certain range of values and they are bounded, or unstable, that is, the variables explode and are unbounded (there are many types of stability but I will discuss only the simplest, intuitive notices). The bounded set to which stable systems tend is called the attractor. It may be a simple equilibrium point, that is a point (a vector whose elements are the values of the variable) at which the system comes to rest. A stable system, once disturbed, will eventually settle back to its attractor. A marble in a basin will find the lowest point, its state of equilibrium. At all other points it will find the lowest point, its state of equilibrium. At all other points it will be in movement.

There are other types of attractor. Cyclical attractors are known as limit cycles, and may describe fluctuating fish populations or business cycles. A recent discovery is that of strange attractors. The solution paths of systems characterized by strange attractors are called chaotic, and are aperiodic yet recurrent and bounded (more on this follows).

Unchecked growth is, of course, unstable. This point cannot be emphasized too much. It growth is unchecked, some variable eventually becomes too large and the system breaks down. Consider a simple equation in continuous time: This equation says that variations in the variable y as time passes are proportional to the level of y, and p is the proportional growth rate. Suppose that y is a measure of the total population of a species, then the equation states that changes in the population at time t are related to the current population level. The equilibrium point is at the value of y for which their is no longer movement, $dy/dt = 0$. For this equation the only equilibrium is at a population value of zero, otherwise the species grows in number. We can find the solution by integration, and letting the initial value, y_0, be positive, we have:

$$Y(t) = y_0 e^{pt}.$$

The solution paths to this simple growth equation are studied in Box 4.1.

In an unstable system with many variables (dimensions), eventually one of the variables reaches a value for which the system collapses. For most situations of interest we assume that there is some mechanism in the system, for example a limiting factor, which prevents the system from total collapse. In population studies this factor can be limited food or the presence of predators. For a very useful and famous example we turn to a simple difference equation formulated in discrete time (variables are measured at equally spaced intervals of time):

$$Y_{t+1} = y_t(\alpha - \mu y_t) = \alpha y_t - \mu y^2_t.$$

Models of this sort are frequent in the study of species populations, and we take as our variable the total number of a given fish species (for more on these types of model in fishery studies see Chapter 5 and Clark, 1985). If $\mu = 0$, the solution is simply $y_t = y_0\alpha^1$ and the fish species grows continuously. If $\mu > 0$, the term $-\mu y^2_t$ is a saturation factor, and the population increases until $y_t = \frac{\alpha - 1}{\mu}$, after which is decreases as can be seen in Figure 4.2. If we set $Y_t = y_t(\mu / \alpha)$ we can write the equation in its standard form:

$$Y_{t+1} = \alpha Y_t (1 - Y_t).$$

This is a single variable, non-linear (quadratic) difference equation which, however, has an entire range of dynamic behaviour depending on the value of α, which indicates the steepness of the hump. In Figure 4.2 we have provided curves with different values of α where the intersection of the 45° line and the curve is the equilibrium point (which can be stable or unstable). If $\alpha \leq 1$, the system has only one equilibrium point (and it is stable) at $y_t = 0$, which means that, whatever the initial fish population, the species eventually becomes extinct (the lower curve). If $\alpha > 1$ the origin is an unstable equilibrium point. For values of $1 < \alpha < 3$ there is a stable, positive equilibrium population to which the species tends and where it remains (at the intersection of the middle curve and the straight line. As the value of α increases from 3 to 4, the dynamics get very interesting. The equilibrium point indicated for the upper curve is unstable, and the population does settle down to an equilibrium size but fluctuates in an unpredictable manner. Then for $\alpha > 4$ the system becomes unstable and the species becomes extinct. This is not the place to go into a complete study of the dynamics involved. However, the implications of recent findings about behaviour inherent even in this simple population example are so important that all of us should at least be vaguely aware of them.

Box 4.1 Exponential growth

In Figure 4.1, solution paths are graphed over a 100-year timespan, beginning at an initial value of $y_0 = 10$ for proportional growth rates $P = 0.01$,....., 0.05. If the rates are yearly, the graph gives the total population for each year. Notice that small differences in the proportional growth rate lead to large differences in the actual number of the species. For instance, starting from a population of 10, after 50 years, the total population is 16 for $p = 0.01$ but 122 for $p = 0.05$. That is, a difference in the proportional growth rate of 0.04 leads to a difference in the population of 106. This characteristic of unstable dynamical systems, that small differences accumulate quickly, is perhaps counterintuitive. Let us look at it from another point of view. Table 4.1 gives the time, in years, that it would take to double the current population at a given proportional growth rate. Taking the previous example, for $p = 0.01$, the population still has not doubled after 50 years, while for $p = 0.05$, the population has doubled approximately 3.5 times in the same period. The formula for calculating the years to double is easy to remember: roughly 70 divided by p per cent.

Table 4.1 : Years to double Figure 4.1 Solution growth paths $y(t) = 10e^{pt}$

Rate	Years
0.01	69.3
0.02	34.7
0.03	23.1
0.04	17.3
0.05	13.9
0.06	11.6
0.07	09.9
0.08	08.7
0.09	07.7
0.10	06.9

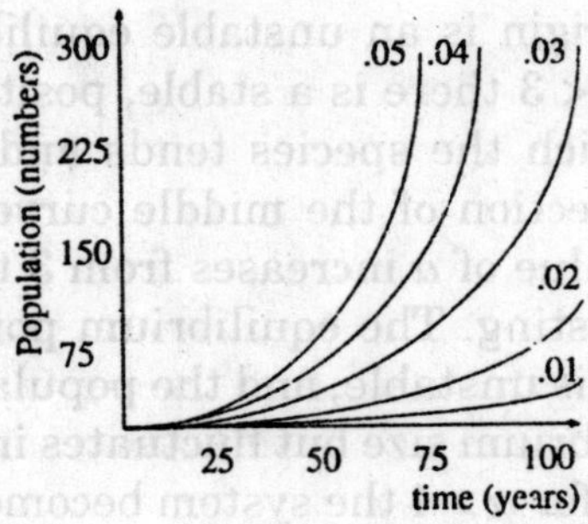

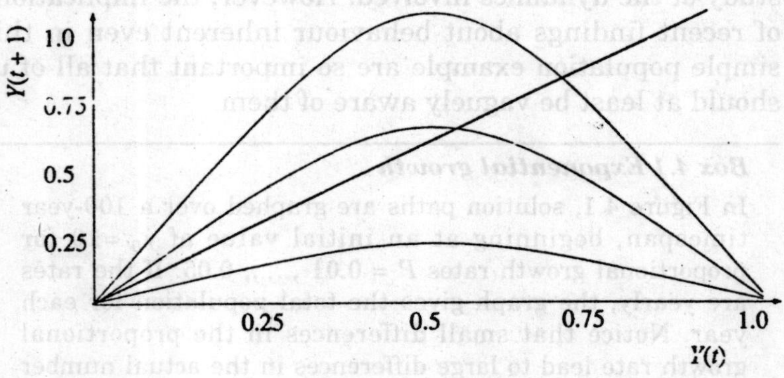

Fig. 4.2 : The logistic equation, α = 0.7 lower, α = 2.3 middle, α = 3.9 upper.

Chaos Theory

In the interesting parameter range $3 < \alpha < 4$, as α is increased, the system first has a series of stable cycle solutions of even-period (2, 4, 8, 16 and so on), then a series of stable odd-period cycles. In Figure 4.3, γ is plotted over 30 time periods for a value of α = 3.53 (the model is discrete but the points have been connected in order to see them better). After some initial transient behaviour, the system settles onto the stable period-4 cycle solution for that parameter value. At α = 3.57 the system actually follows what are called 'chaotic trajectories', that is, solution paths are 'chaotic', they remain bounded and so cycle around, but never repeat the same cycle. The attractors for systems with chaotic trajectories are known as strange or chaotic attractors. Again, their behaviour is interesting because, although it is aperiodic, it is bounded and variables in the system remain within a bounded range.

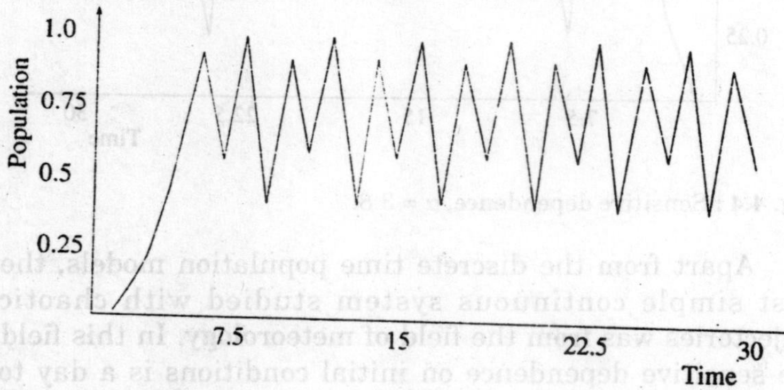

Fig. 4.3 : A period-4 cycle, α = 3.53.

A great fuss has been made about the discovery of chaotic trajectories (which occur only in non-linear systems) due to the property called 'sensitive dependence on initial conditions' (SDIC). This sensitive dependence means that if one starts y_t out at two positions very close to each other, the paths soon diverge. In Figure 4.4, two timepaths are plotted for 30 periods with starting values $y_0 = 0.200$ and $y_0 = 0.205$ for a value of $\alpha = 3.8$. While the two trajectories remain almost identical at first, after five or six time periods they begin to diverge, and after eight time periods the whereabouts of one trajectory gives no information on the whereabouts of the other. Then if our knowledge of the system in the initial period is not infinitely accurate—even though the system is deterministic—we will not be able to predict the future timepath. It is this quality of quick divergence which creates problems for forecasting (as well as for fundamental distinctions between stochastic and deterministic systems).

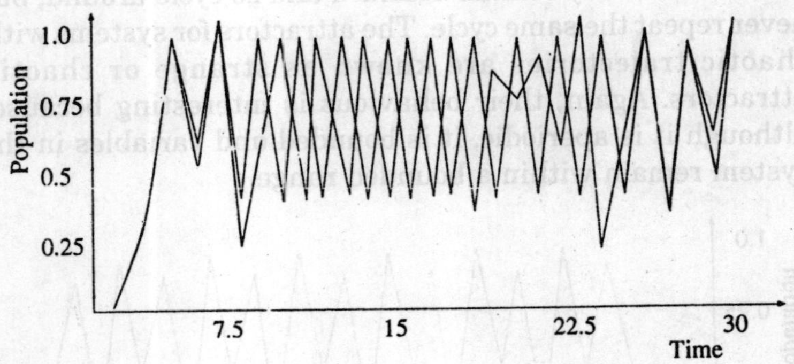

Fig. 4.4 : Sensitive dependence, $\alpha = 3.8$.

Apart from the discrete time population models, the first simple continuous system studied with chaotic trajectories was from the field of meteorology. In this field the sensitive dependence on initial conditions is a day to day fact; except for short-term timescales, prediction is very difficult. This inherent characteristic of meteorological systems has an important influence on the magnitude and location of the consequences of ecological disasters involving emissions. Winds and rains affect not only large-scale catastrophes such as Three Mile Island and Chernobyl, but also the myriad local catastrophes which are part of modern life. The uncertainty introduced into deterministic systems by chaotic behaviour forces the adoption of probabilistic analytical methods and there will be more on this is section 3.

Catastrophe theory

The analysis above is an example of bifurcation theory, which studies how systems change as the parameter values (α in the example) change. Another theory with important implications for the study of dynamics in environmental economics is catastrophe theory, which studies abrupt (discontinuous) changes in a fast system response to smooth, continuous changes in external conditions. More information on catastrophe theory is given in Box 4.2. The exact equational forms necessary to produce catastrophes may not be common to economic analysis, but the qualitative intuitions from catastrophe theory are useful. The mathematical theory of catastrophe theory is well developed; for a readable introduction (by one of the prominent theorists).

I think it is important to note that nonlinear dynamics, of which bifurcation theory and catastrophe theory are parts, instills in those who study it a respect and acceptance of surprises and of the limitations to predictability. This is in direct contradiction to the intuitions of linear dynamics the system either smoothly increase or smoothly declines, and nearby trajectories in stable systems converge through time.

Box 4.2 Catastrophe theory

Consider a boat rolling in a storm. Even as it is continuously rolled to the right by the action of the storm, if a strong enough force is applied it will reach an angle after which the boat capsizes. This type of dynamics is especially dangerous because the system moves along in a reassuring, smooth manner until it reaches a point as which it completely changes behaviour. An example of a particular type of catastrophe surface, called the 'cusp' catastrophe, is given in Figure 4.5. This example is taken from Van Dijk and Nijkamp (1980) and illustrates a possible trajectory of energy use for exhaustible resources as stocks are gradually depleted. The unstable part of the system is shaded, and indicates that the system will not settle there the rest of the surface consists of equilibrium points. The surface is like a sheet of paper with one side folded slightly. A trajectory may begin at point B and move smoothly down the back part of the surface, or flow towards the fold, where it will eventually experience an abrupt change (a fall, a catastrophe).

In the example corresponding to Figure 4.5 the authors consider a situation in which there is an initial increase in an energy stock (k^e), owing to a discovery, followed by the depletion of that stock. It is assumed that there is less inertia (i) and friction in adjustment to an increase in energy use (with rising stocks) than to a decrease in energy use (during depletion). Then energy demand (e^d) changes smoothly (A-B-C) during the increase in resources, but may experience abrupt shifts (C-D-E) during the decrease in resources.

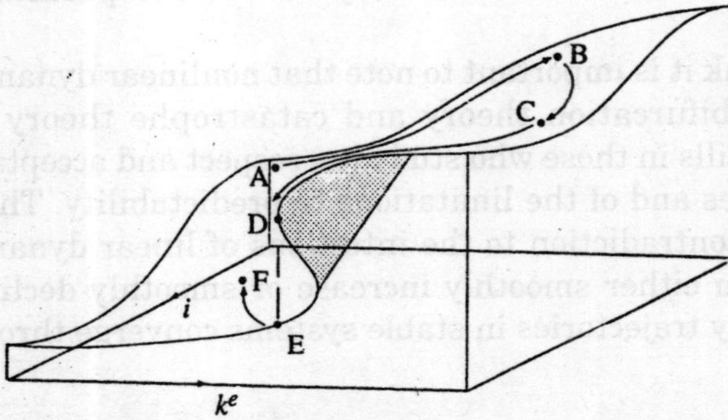

Note: Illustration courtesy of Elsevier Science.

Source: Van Dijk and Nijkamp (1980, Fig. 15).

Fig. 4.5 : Cusp catastrophe surface-exhaustible energy resource example.

Evolutionary systems

A final note on evolutionary systems is in order. So far we have discussed systems whose dynamics are passive, that is, fixed. Adaptive systems, for which the dynamics are modified by interactions (for example, systems with feedback mechanisms), characterize evolutionary processes and cognitive processes such as learning. It is clear that the ecological and economic systems are evolving, even 'co-evolving', at east since the world economic system has reached a level that stresses the environment. From this point of view, economists working on some parts of the environmental-economic system may need to expand their interest in dynamic aspects beyond a focus on, and search for, equilibria and steady states (where the population, output and capital stock grow at the same rate, see section 2.3). If the economic-ecological system is 'co-evolving', it is the evolutionary stability that will prove important for understanding issues of sustainability. The study of dynamics in evolutionary (live) systems is more intricate than the study of dynamics in mechanical systems, and it is at an early stage. In fact, there is no general agreement on which characteristic of a system should indicate its evolutionary stability: persistence, resilience or integrity. Yet evolutionary stability of the economic-ecological system is the essential question if our concern is to allocate resources so that future generation will have similar consumption possibilities (which will require a certain minimum state for the environment).

The issues in this field are still open for discussion and hotly debated hypotheses should appear in the near future. An example follows of a possible policy that evolutionary studies might support regarding the common obligation to biodiversity: give priority to setting aside a specified minimum of the planet where natural systems are allowed to evolve, rather than to handing the next generation an inventory of the exact number of each species existing at a given point in time. The emphasis switches from conserving current species to maintaining the potential for future species to evolve. It must be said that, from a policy point of view, it is much easier to quantify population numbers for a given list of species than to estimate the area necessary to ensure the qualitative characteristic of evolutionary potential. Nevertheless, insights that derive from this worldview should prove useful.

The study of economics as evolutionary systems also leads to important policy implications. Consider a typical evolutionary scenario. Opportunities for a new technology lead to an 'anything goes' competitive among various techniques and entrepreneurs. The field of competitors is gradually reduced until eventually a leader gets enough market share to gain from what are known as increasing returns to scale, that is, there are cost reductions to mass production and distribution. Once this technique has acquired a substantial following, other secondary industries, as well as accustomed clients and consumers, all gain by its continued dominance (up to a point, of course). This is irreversible development; the market does not turn back and choose another keyboard arrangement of the alphabet even if it is more efficient. Thus the economy is path-dependent and it will require a shift to move to another development path. Here public policy must calculate the social benefits and costs before breaking up dominant industries. In the meantime, policies must clearly support those industries and technologies from which society can benefit in the long run.

New ways of modelling evolutionary systems which use the formalism of neural nets and genetic algorithms my eventually offer new tools and a different perspective for viewing the development of the evolving economic-ecological system.

Thresholds and irreversibilities

A closely related issue it that of irreversible dynamics. Some physical processes are obviously irreversible; topsoil once lost from a slope will not find its way back unassisted. (Given the rate at which topsoil is formed, the capacity for the environment to reverse the loss is negligible for most analyses.) Damage may also be irreversible, either because the physical process has been pushed beyond its threshold value and has actually suffered irreversible damage or because, from a technical point of view, the costs of returning the system close to some past or preferred state may be prohibitive. Toxic waste clean-up attempts suggest that, in some cases, irreversible damage in the second sense has been done to parts of the ecosystem.

For many processes, there is a threshold value for some variable which signals the onset of irreversibility, that is, once that value has been surpassed, the process becomes irreversible. The 'forcing' by, say, the dumping of warm water into a lake may cause a bifurcation, a qualitative change in the lake's ecosystem dynamics, as the high temperature forces the lake temperature beyond threshold values for certain species. Irreversible damage may be caused as the ecosystem withers and the lake becomes 'dead'.

Many fish species experience large natural fluctuations in their population. Large marine fishery collapses have often been blamed on commercial fishing activity, but in some cases at least, fishing activity only forces the population closer to the threshold value (below which the species disappears). At that point a series of natural events gives the definitive blow. Actually fish populations are rarely harvested to extinction, but become so sparse that commercial activity is no longer viable.

Of course, other species have disappeared, generally owing to a loss of their natural environment. These pure irreversibilities pose serious problems for the dynamic allocation problem, particularly since damages from future events will be discounted by the policy-maker. If the authority uses a high discount rate, the costs of even catastrophic events 50 or 200 years into the future will seem irrelevant now (see Table 4.2). Since current decisions which may trigger events leading to irreversible damage in the future (such as species loss) must normally compare current costs with future benefits, the choice of discount rate often decides the choice of action. We now turn to a discussion of the rationale and ethics of discounting future utility.

Table 4.2 : Present value of 1 ECU benefit at 50 years and 200 years in the future.

Rate	*50 year horizon*	*200 year horizon*
0.00	1.00	1.00
0.01	0.61	0.14
0.02	0.37	0.019
0.03	0.23	0.0027
0.04	0.14	0.00039
0.05	0.087	0.000058
0.10	0.0085	0.0000000053

Discounting

Ethical issues : Consider a benefit-cost analysis of a project aimed at reducing emissions. The costs will be borne by the current citizenship. Some of the benefits may fall to the same people who bear the costs, but most will go to future generations. How can policy-makers convince current consumers to reduce consumption now through, say, an energy tax? It must be demonstrated that the stream of benefits more than outweighs the current costs, and the stream of benefits will traditionally be discounted so that the further in the future the benefits lie, the smaller are their current values. Is it ethical to discount the well-being of future generation?

The seminal work on the intertemporal allocation of resources is Ramsey's 'A Mathematical Theory of Savings'. For Ramsey, the issue was allocation of national income between saving and consumption, with the idea that investment (not consuming) now, would mean higher consumption opportunities later. He states at the outset, 'that we do not discount later enjoyments in comparison with earlier ones, a practice which is ethically indefensible'.

Most arguments against discounting are based on the issue of intergenerational equity. That is, if we discount the future we are discounting not only our own future utility, but the utility of unborn and unrepresented generations. Rawls (1972) notes that, 'there is no ground for [...] using today's discount of the future rather than the future's discount of today'. Sen (1982) argues that imposing pollution on future generations reduces their 'liberty'. Welfare is typically represented by measures of production capacity, per capita consumer goods and services, and possibly some measure of environmental quality. Therefore, any decision to undertake investment projects whose aim is to eliminate pollution must take into account the 'liberty' which is not usually considered in welfare calculations.

Others have given arguments in favour of discounting, for example, the possibility that our species will become extinct in the future and will not require current sacrifices. A stronger argument is that future generations will probably be richer and will be able to absorb a decline in environmental quality. An implicit assumption for this position is that one can continuously substitute consumption for environmental quality. That is to assume, for instance, that the unhealthy air quality in Milan can be compensated through increased expenditures in, say, interactive television, VCRs and expensive cars. One must also assume that the threshold at which the ecosystem will collapse owing to environmental degradation is higher than the levels that will be reached in any foreseeable future under any imaginable scenario.

If one must discount the future, most environmental economists would prefer a low rate, thus giving more weight to future conditions. Yet Pearce, Barbier and Markandya (1990) warn that a low discount rate can also lead to increased investment, and thus economic growth and further environmental damage, if the interest rate set by the authority is related to the low social discount rate, that is the rate at which society as a whole discounts future consumption (see Box 4.3).

Box 4.3 Social rate of time preference

Time preference refers to the fact that people prefer present consumption to future consumption. The SRTP is some average of individual time preferences and is the rate at which society discount the future in comparison with the present. The idea is that to induce society to sacrifice one unit of consumption today, it must be offered $1 + r$ units next year. Then the future counts less than the present and the rate at which it is discounted is the STRP r. The notion is very abstract, but the value given to the STRP is not just an academic question. Government authorities must base their decision to undertake a given project or agreement on the net benefits of the project. If the benefits accrue during time periods different than those for which the costs are borne, the project evaluation rests on a comparison of the total discounted benefits to the total discounted costs. What is the appropriate rate? This very much depends on the timespan and goal of the project or agreement; a discussion follows.

The fact is that in practice, future well-being is discounted. The issue then becomes: which discount rate should be applied? Useful surveys are Lind (1982, 1990) and Cline (1992) from which the discussion below profits.

Sourcing

In social benefit-cost analysis a traditional approach is to consider the source of the funds for a given project and use the rate appropriate for that source. Public sector infrastructure projects such as highways, bridges and railways are typically financed from the national budget. The appropriate rate is then the rate of return on alternative public sector projects. Some environmental projects, such as abatement schemes, may be of this type. For other projects, such as taxes on petrol or energy consumption, costs will be borne by the general economy, for which the share of consumption is around 80 per cent and the share of investment around 20 per cent. The appropriate discount rate for consumption (how much the value of a unit of future consumption should be diminished in comparison with the value of a unit of current consumption) must be the 'social rate of time preference' (SRTP; see Box 4.3). Owing to taxes on capital gains and imperfections in the capital market, the SRTP is generally held to be less than the investment rate of return. The appropriate rate for the energy tax evaluation, then, is an average of the two, each weighted by its share in the economy. An example is given below.

Suppose that a carbon tax is proposed to reduce emissions. For reasons of simplicity, assume that the costs will be borne by the current generation and benefits accrue to future generations up to 200 years into the future. If the SRTP is around 3 per cent, the real investment rate of return around 10 per cent, and the consumption and investment shares 80 and 20 per cent respectively, then the correct discount rate to apply will be:

$$(0.03)\ (0.80)+(0.10)\ (0.20) = 0.044.$$

Thus by weighting the discount rate according to the source, the correct discount rate (4.4 per cent) is much lower than the investment rate of return (10 per cent), and a decision to enact the policy is more likely to be made. The final comparison must be between current costs and the discounted stream (at sourced rate 4.4 per cent) of benefits accruing from the first benefits (say at t) through 200 years into the future. If:

$$Costs(t) - \int_{t}^{200} Benefits(t).e^{-0.044t}\, dt < 0,$$

the project is approved. Although the present value of one US$ spent 200 years from now is small at 4.4 per cent (0.00018 cents), it is significantly larger than the value at 10 per cent (see Table 4.2). It is more probable that the sum of all the discounted benefits is sufficient to satisfy the condition above if the lower, weighted rate is used.

Consumption-equivalence

Current practice is to convert the investment effects 20 per cent in the example above, into consumption equivalents and discount the total at the SRTP. The conversion is accomplished by computing a 'shadow price' (see Box 4.4) of capital which is multiplied by the amount of private capital displaced by the public sector investment. This gives the 'true opportunity cost in terms of consumption'. A study which uses the consumption-equivalence approach is the one by Kolb and Scheraga (1990). The authors consider various benefit streams for anti-pollution equipment with an expected 20-year life and find that the approach gives higher benefit-cost ratios than straightforward discounting. Moreover, the difference between the two is more pronounced the further in the future the benefits are, or the longer the gap between costs and benefits. Naturally, the 10 per cent rate recommended by the Office of Management and Budget (OMB, see below) would eliminate any long-term, and most middle-term, projects.

Box 4.4 Shadow price of capital

In general, shadow prices are distinguished from market prices in that they reflect all of the indirect consequences of an action. The social shadow price reflects all of the social costs and benefits issuing directly or indirectly from a project. Thus the 'shadow price' of capital (machinery, buildings, and so on) is not the sale price but the value imputed from capital productivity over the lifespan of the capital good. Since by investing one increases the amount of capital and thus the capacity to produce, additional consumption goods will be available in the future, and these must be taken into account when calculating the shadow price. These future additional consumption possibilities due to the increase of capital at time *t* are discounted back to *t* and summed. The shadow price is a relative price and in the optimal growth problem is expressed in (usually current) units of the consumption good (example: one small abatement investment with a lifespan of 15 years is equivalent to 10^9 bread loaves today).

The debate continues over just how the shadow price should be calculated. Although various authors use different techniques, the range of values is limited: approximately 1.0, Bradford (1975); 1.5, Cline (1992); 2.94, Kolb and Scherage (1990); 3-5, Lind (1983).

The appropriate rate

The sensitivity of any benefit-cost calculation to the discount rate should not be underestimated. In Table 4.2, the present values of a future ECU benefit 50 and 200 years into the future are calculated for various discount rates compounded annually. The 50-year future timescale is relevant for issues such as resource exhaustion (including some mineral deposits, fisheries, forests and so on), alternative energy projects, agricultural management incentives and pollution abatement schemes. One example:

the Environmental Protocol signed in 1991 under the Antarctic Treaty System effectively enacted a 50-year moratorium on mineral resource activity. The 200-year timescale is relevant for such environmental issues as global warming, fossil fuel resource exhaustion, research and development for alternative energy sources, and biodiversity.

For a look at current practice consider the US guideline rates for evaluation of government proposals during the 1970s and 1980s: the Office of Management and Budget (OMB -executive branch) suggested a real discount rate of 10 per cent; the General Accounting Office suggested the Treasury borrowing rate for maturities with length near the life of the project (the average real interest rate on long-term US Treasury bonds (1956-91) was around 2.6 per cent); and the Congressional Budget Office suggested a real discount rate of 2 per cent. The difference between the high rate guideline of the executive branch (the OMB) and the lower rates used by congressional offices over the period under consideration probably reflects the difference in political philosophy between Republicans (who controlled the executive branch) and Democrats (who controlled Congress). The divergence indicates the power of the discount rate to decide for an active or inactive investment policy.

When actually applying SRTP for long-term projects analysts have generally looked observable market rates, such as the real after-tax rate of return. Lind (1982) adopts 1 per cent as an approximation of the real rate of time preference associated with a safe asset (US Treasury bills) and 2 per cent as an approximation for safe long-term assets (long-term US government bonds). Kolb and Scheraga (1990) adopt 3 per cent as midway between the 1 per cent average real after-tax rate of return on three-month US Treasury bills and the 5 per cent average return on stocks.

Cline suggests an alternative method for deriving the SRTP based on the utility function's concavity. Economists usually assume that the function which translates consumption into utility is concave. That is, utility from a further unit of consumption is always positive, but the amount declines as the level of consumption increases. (Consider ice cream: the first scoop is wonderful, the fifth less so.) There is a natural discounting of the same person's future utility if that person expects to have higher levels of consumption in the future. Such has generally been the case during the short histories of market economies, at least for the property-owning classes. It is not clear, however, that productivity will increase forever. Even if small increases can be assumed for several centuries, world events may require a greater sharing of consumption goods, and this redistribution may in fact imply a reduction in future consumption levels of those who are currently privileged.

In any case, given the concavity assumption and taking a typical functional form for utility, Cline establishes that the SRTP should equal the growth rate of per capita income (using the 'golden rule' notion discussed below), which has averaged about 1.7 per cent annually over the past century in the USA. If one considers the global per capita income growth, of course, the figure is lower and in the very long run Cline concludes that an SRTP of the order of 1 per cent annually would seem appropriate. For a more general class of utility functions, the SRTP would be 1.5 times the growth rate of per capita consumption, or closer to 1.5 per cent.

Many economists will not be happy with such low social discount rates, but given that the appropriate discount rate will very according to the project at hand and the Cline's concern is global warming, I think he has a strong argument. He quite rightly concludes that under the OMB-recommended 10 per cent rate, the best greenhouse policy will almost certainly be to do nothing.

Optimal Growth Theory

In growth theory the economist is usually concerned with finding the growth path which leads the economy to an equilibrium position, known as the steady state (see Box 4.5). The 'optimal solution path', for which per capita consumption is maximized, is derived from the hypotheses of the model. The golden rule (do unto others as you would have them do unto you) in growth theory is to provide the next generation with enough capital investment to keep per capita consumption (production) at its maximum level. Basically this requires that enough from each generation is saved and invested to maintain a constant ratio between total capital and the total labour force (the capital-labour ratio). Since output per worker is a function of capital per worker, as long as the capital—labour ratio is constant, the per capita output ratio is too. So in the steady state, output per worker and capital per worker constant, but the output and capital grow at the same rate as the population. Then under usual assumptions about production functions, the optimal steady state (maximizing per capita consumption) is that for which the rate of population (and output and capital) growth is equal to the marginal product of capital. The 'marginal product of capital' is the extra units of output that can be produced by an extra unit of capital, and is an indication of the productivity of the capital stock.

Box 4.5 The steady state

The steady state is that solution to the system for which the population, total production and the capital stock all increase through time at the same proportional growth rate. It does not refer to a fixed equilibrium point as discussed in section 2.1. All three variables are growing in a moving equilibrium which ensures equality between the rates of proportional growth. For the simple differential equation discussed in section 2.1:

$$\dot{y} = \rho \, . \, y,$$

where $\dot{y} = dy/dt$, the proportional growth rate $\dot{y}/y = p$. If y represents production, then in the steady state both capital stock and the rate ρ. This would be a happy coincidence because output, and therefore consumption, per person would remain constant.

Of course, we are not in a steady state. Output per worker has steadily increased through the short history of capitalism. But the golden rule does provide an indication of an optimal and economically sustainable state which is also morally acceptable.

The dynamic optimization extension of growth theory is at once more realistic and more complicated. I have included it because much of the theoretical discussion on environmental economics derives from applications of this technique. A critique follows the presentation of the model.

The problem

A typical, abstract framework that economists use for studying the optimal dynamic allocation of resources (capital, lobour, natural resources) through time takes as its basis the maximization of social welfare. Social welfare is the aggregation of direct utility from consumption and the consumption-equivalent effects of investment (see consumption-equivalence in section 2.2). The agregation is assumed to include both the well-being of people living at different points in time and the well-being of different people living in the same period. This assumption poses limits to the types of function that can be used to represent welfare. But on a more abstract level, there is also a strong underlying hypothesis of a seamless social fabric; the welfare function is the same everywhere and throughout time. This is not the place to discuss the appropriateness of the intragenerational homogeneity assumption: I only note that an outsider to the field may be surprised at how rarely the issue of welfare distribution among people living in the same period arises in this optimization framework.

In a dynamic context, even the intergenerational constancy of the social welfare function is certainly non-trivial. As world population comes to exceed the carrying capacity of the planet, one would expect environmental quality to be given priority over consumption of non-basic goods in preferences. No amount of high-tech video and audio equipment will be able to compensate for the scarcity of fresh water, for example. One may even argue that the possibility for global sustainable development depends very much on changes of attitudes among the members of the developed countries. On the other hand, regional conflicts and economics crises may lead to less sensitivity to environmental issues. In most, if not all, cases, the social welfare function, *W*, is assumed to be constant over the time horizon.

In general, it is not obvious how much of the future one should take into account for the maximization process, and in order to avoid the problem of choosing a date, economists often consider the infinite future. Since the future is exponentially less important, distant welfare will have little effect on the total anyway. Then, the maximization of social welfare over an infinite time horizon can be formalized as:

$$max\, \omega(t) = \int_0^{\infty} e^{-rt}\, W(\mathbf{x}(t))dt,$$

where boldface indicates a vector of variables which contribute to welfare: $\mathbf{x}(t) = (x_1(t), x_2(t),\ldots, x_n(t))$ such as agricultural products and household appliances as well as environmental goods such as clean air, drinkable water and so on. Then the idea is to maximize welfare from here to eternity, discounting future welfare at the SRTP, *r*, to give total welfare in present value terms. Once a value for the SRTP has been agreed (see Box 4.3 and the discussion in section 2.2), the problem is one of dynamic optimization given the constraints. The constraints vary according to the problem at hand. Some typical examples are: quantities must be non-negative, the total of consumption and the various forms of investment cannot be larger than production, and environmental quality remains above a threshold value. The constraints are introduced into the problem by way of Lagrangian multipliers, and the time paths for the shadow prices of capital and the environment are derived from the constraints. The shadow prices, again, are the imputed values of the variable in (usually current) terms of the consumption good. The shadow price of capital is discussed in Box 4.4. The shadow price of the environment must take into account the assimilative capacity of the environment (more on this below).

This is the general form, and any number of models can be generated depending on the assumptions made regarding utility and production.

Guidelines

Necessary conditions for an optimal path provide the authority with guidelines for action. An important general conclusion is the *Keynes-Ramsey rule* which implies that consumption increases, decreases or remains constant, depending on whether or not the marginal product of capital per capita is greater than, less than or equal to the SRTP. The golden rule condition, the marginal product of capital per capita, is equivalent to the growth rate of the population, ensures that consumption per worker is constant and is constant at the maximum possible level, given the assumptions. This is the moral code that each generation leaves behind enough capital per worker to permit the next generation the same level of consumption. In the case of dynamic optimization with discounting, the modified golden rule is that (net) marginal product of capital per capita must equal the discount rate. For a growing population this means:

$$MPCapital = SRTP + n, \text{ rather than } MPCapital = n,$$

where n is the growth rate of the population. A typical optimal growth model with infinite horizon will take the population as constant ($n = 0$). However, there will be a negative externality to production in that emissions are produced, and the modified golden rule is that the marginal product of capital per capita net of the externality effect must equal the social rate of discount.

Then the authority should lower the consumption level, that is, increase savings and invest more, if the net marginal product of capital per capita (again, the contribution of the last capital employed) is greater than the rate of time preference. If people are discounting very little and capital is highly productive, it pays to switch some national income from consumption to investment so as to increase consumption later.

Secondly, this switching will be made all the more probable, the higher the rate of substitution between consumption today and consumption in the future. If people are living only for today, the condition for foregoing current consumption will be more difficult to reach.

Environmental issues present two basic problems for optimization—one problem is finite resources, the other is finite assimilative capacity of the environment. Taking the first, it would seem difficult to square intergenerational equity, that is, a constant consumption level, over an infinite time horizon if exhaustible resources are being bound up in produced goods from which 100 per cent recycling is impossible to attain. The solution, known as the *Hartwick rule* (see Hartwick, 1977), states that per capita consumption can be held constant with declining stocks of resources only if the rents (the difference between extraction costs and the price) arising from the efficient use of these resources are invested in man-made capital. The idea is that the optimal policy is to maintain total capital stock at its existing level, but since the natural capital stock is declining, the stock of man-made capital must be increasing. The rule requires sufficient substitutability between exhaustible and reproducible stocks for productive purposes. In fact, the power function generally used to represent production (known as the Cobb-Douglas production function) permits such substitution and aggregate output can be maintained constant in per capita terms.

This optimistic conclusion rests on smooth continuous transitions but real-world economies might experience discontinuities. For example, technological innovation may not be in step with resource depletion for a number of reasons: (*i*) subsidies and protection of current products and processes; (*ii*) the irreversibility of investment (internal combustion automobiles are not easily converted into solar-powered vehicles); and (*iii*) the prohibitive costs of research and development of alternative energy sources and machines able to use them. Policy-making bodies should be aware of these difficulties and (slowly) cut subsidies to depletable resource industries and provide research grants and other support to the infant industries of alternative energy sources.

Various models exist for the second problem. A representative model would include an environmental good (natural capital) and two sectors, the usual production sector and a second devoted to partial abatement of the pollution arising from the first sector. There is a maximum tolerable stock of pollution; changes in pollution stock are equal to emissions minus abated pollution minus the given assimilative capacity of the system. The authority controls the distribution of national income between abatement and production activities. The social shadow prices of capital and the environment (see Box. 4.6) are derived, and the social planner imposes the shadow price of the environment as a tax or unit waste charge.

Box 4.6 Social shadow price of the environment

While many assets have market prices, environmental assets often must be imputed a value. The production process has two outputs—the product and emissions. It will then damage the environment whenever pollution levels surpass the assimilative capacity of the environment. These future environmental damages caused by increased emissions (production) at time *t* are discounted back to *t* and summed to determine the shadow price. The social shadow price of the environment is then the total social cost of increasing emissions at *t*; it is therefore the value which the government should use for a tax on emissions.

If there is a steady-state equilibrium where the state of the environment is also constant, net emissions must equal the assimilative capacity of the environment. That is, all emissions are abated except an amount equivalent to the flow normally assimilated by the environment. Further, the social discount rate must equal the social value of the net marginal productivity of capital (per capita). It can be shown that increasing the SRTP lowers the steady-state level of both capital and environmental quality with typical assumptions. This is a fairly intuitive conclusion: by giving more weight to current consumption, environmental quality will be lower and the stock of capital will be smaller in the steady-state equilibrium.

Difficulties in the method

For the problem above if the SRTP is sufficiently small, the steady-state equilibrium is a saddle point. A saddle point is a typical solution to a problem in classical mechanics from which the optimal growth framework is taken. Imagine a mountain range with a narrow ridge between two peaks. Then the only local stable position is between the two peaks on the lowest point of the narrow ridge, the saddle point. A saddle point has two solution paths which converge to it, but all other solution paths are unstable, so that nearly all initial conditions lead to instability and any disturbance to the system carries it away from the saddle point. Essentially, this dynamic property puts heavy demands on the information the authority must have in order to guide the system to that time path leading to the steady state. Furthermore, the path to steady state may in fact be dominated by other solution paths which have a higher present value of welfare. Also, the fact that optimal paths exist does not guarantee their reachability from all initial points. There are a number of conditions that one can use to guarantee sufficiency which basically establish the 'normality' of the problem that is, the invertibility of the matrix of shadow prices which ensures that there is a control for every resource. Therefore it is necessary to assume that the authority not only has all of the necessary information, but also all of the policy instruments necessary to put the economy on one of these stable paths and to move it back should disturbances start it off on a divergent trajectory.

Alternatively, one must assume that the market mechanism incorporates social attitudes and somehow sets prices to include shadow prices in such a way as to arrive at, and stay on, the stable path. Furthermore, to assert that buyers of the resource actually face these optimal prices requires a rather far-fetched assumption that there exist spot and futures markets for every resource and capital good, and for all possible contingencies, from the present to infinity

Other uses of control theory

Optimal control is also used to solve dynamic optimization problems in resource management: optimal harvest for renewables and optimal depletion for exhaustible resources. The optimal use of a renewable resource such as a fishery or a forest requires that the harvest is constrained to be no greater than the sustainable yield. 'Sustainable yield' is defined as the amount that can be harvested while maintaining a fixed population.

The price of an exhaustible resource will clearly rise as the resource becomes scarcer unless preferences change. Then the optimal depletion rate (known as *Hotelling's rule* from Hotelling, 1931) is that which equates the rate of price increase of the resource with the interest rate. This is a very general rule in economics, and is just a description of the common sense rule of thumb used by a property owner, say the owner of a copper mine. Call the proportional rate of increase (see Box 4.5) in income from copper sales received by the owner his profit rate. If that rate is less than the interest rate obtainable by selling off the mine and investing the money, the owner will deplete the resource. This view reflects the notion that in a competitive economy, all investment opportunities must yield equivalent rates of return. Although this might not always be the case, one expects that in a competitive market profit opportunities will soon be exploited and disappear (the efficient market hypothesis). On the one hand, if the price is too low, investors will deplete the resource in currently existing operations; if the price is too high, on the other hand, new operations will be undertaken.

Concluding Remarks

What can the environmental policy-maker learn from the study of dynamics? There is no easy prescriptive conclusion here; what is to be gained is in the depth of understanding of how systems change qualitatively and develop through time. Policy implications will follow logically from a far-reaching vision informed by an awareness of the possibilities of discontinuities, abrupt changes and the rapid development of unforeseen consequences in the complex, integrated ecological-economic system. In order to resolve the series of environmental crises as they occur, leaders should have a firm foundation of the following notions and concepts.

The policy-maker should take away a sobering sense of the accumulation of effects through time. Behaviour which seems sustainable when viewed with a short-run lens my, in fact, be unstable even in the medium run.

Unchecked growth is unstable and unsustainable, no species is exempt from this fact. Box 4.1 is a good reminder of the rapidity with which exponential growth increases absolute levels. Recall that an annual proportional increase of, say, 3 per cent means a doubling of whatever variable (be it population, per capita GNP or emissions, per capita consumption of energy or groundwater and so on) every $70/3 \approx 23$ years. Contrast this speed to that with which international decision-making bodies become aware of problems, commission studies and negotiate compromises. One may even be optimistic about the potential of science and technology to resolve environmental problems as they arise, and yet have serious doubts regarding the social, political and psychological capabilities of the species to confront the problems with the necessary speed and flexibility. It should not be forgotten that deep into the Great Depression eminent economists were preoccupied with the risk of inflation.

A policy-maker should not rely on the linear intuition that continually increasing or decreasing a variable will cause constant increases or decreases in another variable. He should keep the picture in Box 4.2 in mind, and worry abut how close to threshold values a proposal will push the system involved. He should pay special attention to how the possibilities of irreversible damage are treated in the analysis.

In particular, if probable damage is in the long run and discounting is applied, irreversible damage may not be obvious in the social benefit-cost analysis. Great care must be taken in deciding the discount rate: it will often decide the issue. Table 4.2 clearly indicates how little the distant future (two centuries away) affects today's decisions under almost any SRTP.

For short - and medium-run projects, the sourcing of the effects (either investment or consumption), will provide a more realistic value for the discount rate.

The price that society imputes to environmental goods, that is, the shadow price, must include all of the consequences, direct and indirect, of the use of that good. The market either gives no value to such goods or it undervalues them because, until recently, there have been zero or low charges for waste disposal. In fact, most economic theory has been based on the 'free disposal' assumption, reflecting the reality of industrial production. Now that average consumption in modern economies has reached a level which threatens the global ecosystem, such policies are no longer viable. Shadow prices are not only important tools to raise funds for environmental projects, but very important educational tools to help consumers adapt to the real costs of what used to be called 'free goods'.

The notion of a shadow price is also important for non-environmental goods. For instance, German law has indirectly increased the price of consumer durables (for example, household appliances) by forcing companies to deal with the lifetime consequences of such goods. Consumers return the discarded appliance to the manufacturer, who must dispose of it. The market price must increase to approximate more closely the shadow price. Again, the educational value of such policies is unestimable, even concerned environmentalists may be surprised at the true costs of our convenience and comfort.

Empirical work on shadow prices will contribute to the identification of sustainable practices, and should be supported. This knowledge should eventually lead to the formulation of feasible development models which can be proposed to developing countries.

From economic theory (and economic history) it is clear that a crucial role will be played by technological innovation in ensuring smooth transitions to different energy sources and different consumer products. Policy-makers must be attuned to these exigencies and find the courage to reduce protectionism and subsidies to those products and processes that are not sustainable. More institutional support will be required to help new technologies in the research and development stage. Further, it will be necessary to help the infant technologies acquire a market share large enough to realize the benefits of increasing returns to scale which will, in turn, translate into lower consumer costs. Without such assistance, it will be difficult for such technologies as photovoltaics to become competitive.

Finally, from growth theory we have the golden rule condition which ensures that one generation leaves enough capital to the next so as to maintain consumption levels: the merginal product of capital per capita must equal the proportional rate of Growth of the population. In the dynamic optimization version, the modified golden rule requires that the marginal product of capital per capita equals the sum of the SRTP and the proportional growth rate. If environmental quality is taken into account the modified golden rule still holds, but the marginal product must be net of the externality effect caused by the emissions from production, and generally the population is considered constant.

The authority must then adjust consumption and investment levels according to the SRTP. If people are discounting the future heavily and capital is not very productive, more of national income will be dedicated to consumption.

In an economy with emissions from production and abatement of emissions, a steady with constant environmental quality requires that total net emissions are no more than the assimilative capacity the environment can handle. Moreover, an increase in the SRTP lowers the steady-state level of environmental quality and capital stock.

It will be obvious that certainty had been assumed throughout the discussion above. Let us now turn to the complications that uncertainties of various sources present to the analysis of dynamic problems in environmental economics.

Uncertainty

Sources and Modelling

In order to understand environmental economic issues, be it for prediction purposes of policy analysis, economists must formulate models. Unfortunately, the enterprise of modelling these complex relations is subject to a great deal of uncertainty. Of course, limits to knowledge constrain any effort to model real-world phenomena, that is, any effort to reduce the infinite-variable complexity of the system to a formalized system of relatively few variables and equations. In addition, there is the uncertainty arising from the interdisciplinary links that vary in environmental economics from the private profit maximizing extraction of minerals to the essential relations between global economy and global ecology. Thus both social and physical processes are involved and must be modelled at many scale levels and across many time horizons. We now turn to an overview and a breakdown of the sources of uncertainty, including examples drawn from resource and environmental economics. Parallel examples will be given in cases where a distinction between the uncertainty in social sciences and uncertainty in the physical and natural sciences is useful.

Measurement problems

Physical aspects will be measured to an accuracy given by the measuring device, and this is true throughout environmental sciences. In economics, one is also confronted with measurement problems of tangible objects such as quantities of output, emissions and capital. Furthermore, these quantities are always some sort of aggregation and one must find a method of converting quantities of different types of output or capital into a common unit of comparison. This is usually accomplished through a monetary valuation. Imprecision can result, for example, from the method of aggregation or from quickly changing prices.

In addition, the economist must consider intangible aspects whose definitions are not exact such as utility, as well as assign numbers to parameters whose values can only be guessed, such as the SRTP (see Box 4.3) and the coefficients of utility functions.

Another category includes errors resulting from classical inference an bias due to non-representative samples, as well as modelling errors such as the omission of important variables.

From dynamical systems analysis we have yet another type of measurement problem: temporal processes may be subject to complicated dynamics. As discussed in section 2.1 (Chaos theory), the nonlinear relations in a system may lead to a stable solution of a 'strange attractor' whose solution paths are chaotic. In this case, even if the system were entirely deterministic and low dimensional, the time series of the variables would appear to be stochastic and the characteristics of the system may be best described in probabilistic terms. The source of this type of uncertainty is the finite precision with which calculations are performed. That is, given infinite precision, one could track and predict the course of the variable exactly. As it is, calculation errors are exponentially compounded in time due to the dynamics of the system, and any but very short-run predictions of a given path will be mistaken. This sensitive dependency on initial conditions is also called 'intrinsic stochasticity', to distinguish it from stochasticity due to external perturbations. It should be noted that although exact long-run predictions are impossible, one may have a great deal of knowledge about the attractor itself, that is, average values, average divergence rates and so on and, moreover, some parts of the attractor may be quite predictable.

On the other hand, a deterministic system may be so complicated because of the high number of important variables, its high dimensionality, that a probabilistic understanding is the only one reasonable. Here again, uncertainty is ascribed to the system from the point of view of the theorist who is trying to understand it.

System stochasticity

The second type of uncertainty refers to the stochasticity in the system itself. Processes may necessarily be described by probabilistic techniques due to microlevel fluctuations or randomness at a fundamental level, such as is found in quantum mechanics. The consequences of fundamental uncertainty for predicting from dynamical systems are, in one sense, similar to those deriving from chaotic bahaviour; there are unsurmountable limits to our precision which compromise our ability to forecast.

Stochasticity may also derive from the fact that few real-world systems are truly closed to exogenous disturbances. The implicit assumption of deterministic analysis is that the surrounding environment is either constant or does not influence the system, or that the system of interest adjusts to the average properties of a non-constant environment. It is the latter hypothesis, generally, that underlies both the habit of scientists to treat the system as if it were closed, and the habit of some to reintroduce the lost environmental effects through the addition of a noise process. The assumption that systems react to the environment's average properties holds up well as long as the models of the system are linear. However, when a nonlinear system is coupled to a stochastic process, the local stability characteristics may well be altered. Suppose a forest ecosystem is in an equilibrium cycle, but that owing to the urbanization of the surrounding area the forest is impacted each year with greater force. It is quite likely that, after a transition period, the forest settles to a new equilibrium which may have very different properties than the original ecosystem. For evolutionary processes, the entire development path may be shifted, and may be shifted to such a point that there is no path back to the previous state, that is, the disturbance may cause irreversible change.

Modelling uncertainty

One very general method of dealing with uncertainty in theoretical analyses is to introduce random disturbances into the deterministic systems. The adjective 'random' refers to the fact that the disturbances take on alternative values according to some probability distribution, and a random or stochastic process has structural characteristics such as an average (mean) and an average deviation from the mean. Most dynamic analyses which assume a random process (also called a noise process) employ a covariance stationary process, that is, with finite constant mean and autocorrelation (between two points in time) strictly a function of the time separating the observations. While this might seem a strong assumption, many modellers use an even stricter hypothesis for the autocorrelation, that is, an obervation is uncorrelated to any other taken at a different point in time. This noise process is referred to as a no-memory process or while noise process.

Single equations

Another important factor in formalizing uncertainty is how disturbances impact the system. Consider the simple one-variable discrete time equation of section 2.1 and let y_t represent the total number of a given fish species at time t:

$$Y_{t+1} = \alpha y_t(1 - Y_t).$$

Of course, the fish population is not just a function of its past values. Cell ε_t the sum at time t of all types of disturbance or effects not represented by past values. If this aggregate seems only to shift around the level of y_t in a random way, then the random variable is simply added, and the noise process is additive:

$$Y_{t+1} = \alpha Y_t(1 - Y_t) + \varepsilon_t$$

On the other hand, it may seem that the parameter itself varies randomly in time, so that the growth rate is subject to disturbances at t, α_t is modelled as a random variable, and the noise process is multiplicative:

$$Y_{t+1} = \alpha_t Y_t (1 - Y_t).$$

Finally, both additive and multiplicative noise may be assumed:

$$Y_{t+1} = \alpha_t Y_t(1 - Y_t) + \varepsilon_t$$

The most common probability distributions assumed for the noise process are the uniform and normal (Gaussian) distributions. The case for the former, where uniform probability is given to values over the range (support) specified, is that the random values are finite, and the maximum and minimum values are known. Such is the case of a die: each side has probability 1/6 of coming up. In the Gaussian distribution, random numbers are concentrated around a mean in a bell-shape distribution. However the 'tails', the parts of the bell that extend out, are infinite, so that there is a (small) probability of a very large positive or negative number coming up. Intuitively it would seem that any system becomes unstable if a large enough disturbance crashes into it, and one may prefer to restrict the range of the values used in modelling the noise. This is especially true when negative values are unacceptable, for example, if they imply negative resource stock values (negative fish and trees) or negative prices (the consumer is paid for taking the good). The Gaussian no-memory process is often used, in part due to its tractability, but also because it seems to fit out notion of what a 'normal' distribution should look like. It should be noted, however, that truncating the distribution has negative effects on the mathematical simplicity of using the Gaussian distribution.

Large systems

Analytical result for nonlinear models coupled to noise processes are rare and generally limited to systems with one or two variables. In order to understand systems for which no analytical solutions exist, one must turn to numerical simulations. In the case of large systems with uncertainty, the dynamic behaviour of the systems, and their reactions to specific policies, will not be single-valued time paths. Information from many runs of the model for a given policy will be represented in a distribution of outcomes. The decision-making body can judge the effectiveness of the policy for a set of objectives stated in probabilistic terms, such as cumulative probability of desired outcome, maximum permissible percentage of negative outcomes, maximum or minimum expected values and so on. In the next section we turn to a case study which illustrates the difficulties involved. A second case study demonstrates how uncertainty can be studied by looking at the changes in outcome due to a change in parameter value, and how differences in perceiving uncertainty can be used to advantage in negotiating uncertain futures.

Case Studies

Ecosystem management of marine resources : The Commission for the Conservation of Antarctic Marine Living Resources (henceforth, the Commission) was established in 1981 in response to the overexploitation of many of the region's biological resources. The following description of the Commission and its work relies on Nicol and de la Mara (1993).

One of the responsibilities of the Commission was to manage the krill fishery on which most of the threatened species of the Southern Ocean feed. This effort has led to an attempt to manage an entire ecosystem under conditions of great uncertainty and serves as an interesting case study of the problems involved.

The Commission opted for a management style called 'feedback management' over the commonly applied reactive management, as the latter (that is, take action when it is obviously necessary) had led to the collapse of a number of marine fisheries. Feedback management uses control theory, discussed in a cursory manner in section 2.3 regarding simple economic models, and discussed in section 3.3 for those models under uncertainty. In the present case the adjective 'feedback' describes the essential interactive character of the approach. The authority develops an integrated model of the ecosystem dynamics. This model serves of identify critical variables and their threshold values, as well as a basis for numerical simulations under hypotheses of alternative policies. The system's attributes that are to be controlled are then monitored and measurements are compared with target values (objectives). If the observed and objective values differ, the authority activates some mechanism over which it has control and which influences the level (or growth rate) of the relevant variable. The consequences of this action on the system are then observed in new measurements, so that management has continuous feedback on its performance.

The krill fishery raises a number of problems related to modelling uncertainty. A krill population model requires the following minimum information (the sources of uncertainty connected with the information type are in parentheses): an estimate of the initial abundance (measurement errors); the natural rate of mortality including predation (lack of scientific knowledge); individual krill model of weight gain during lifetime (lack of available scientific knowledge); and juvenile recruitment, that is, the number of young actually joining the adult population (natural fluctuations).

As is often the case for marine resources, the krill population experiences natural fluctuations from year to year due to the great variability in recruitment. In turn, this variability is due to the stochastic nature of weather conditions, variation in ocean currents and so on. In practice, this uncertainty deriving from disturbances is irreducible: 'Even with the best data, it will always be impossible to attach a number to the dynamic krill population'.

It would be unrealistic in these conditions to base decisions on a single point estimate of the model. Rather the model is used to predict a distribution of the probable annual population. Uncertainty in modelling the population derives essentially from two classes: modelling uncertainty regarding parameter values (mortality rate and growth rate in weight) and initial conditions (current total biomass), and uncertainty due to disturbances to the system (the natural fluctuations in recruitment). Under the assumption that the crucial uncertainty is the annual number of young krill, numerical simulations can be run for the given model using a different random pattern for that variable at each trial. By running hundreds of simulations a distribution for the biomass is obtained. A sample distribution of krill biomass is given for the Commission's model in Figure 4.6.

The effects of policies can be studied by introducing human predators into the model, and for a given catch rate hundreds of simulations are run. In Figure 4.7 a sample distribution with fishing is provided for a given constant catch rate. A threshold value for biomass is indicated in the distribution with fishing and the probability that this particular catch rate will endanger the population is the amount shaded to the left of the critical value. Once a maximum acceptable risk is decided, the appropriate catch rate can be selected. Parameter uncertainty may be studied by using a constellation of parameter values which leads to extreme cases, if it is possible to determine such a constellation. Otherwise, a statistical technique known as

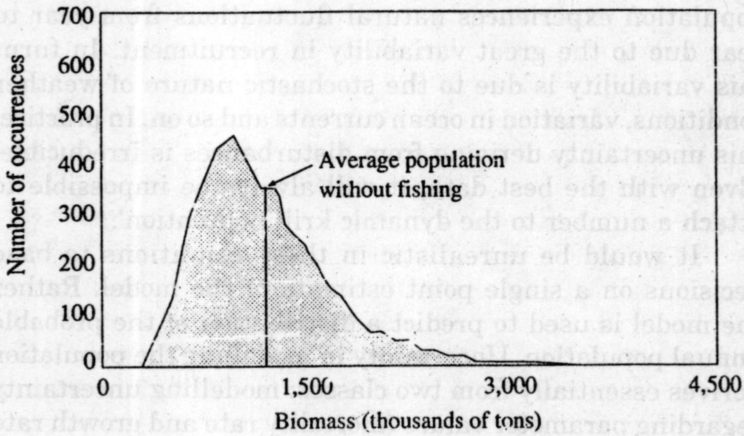

Fig. 4.6 : Biomass distribution.

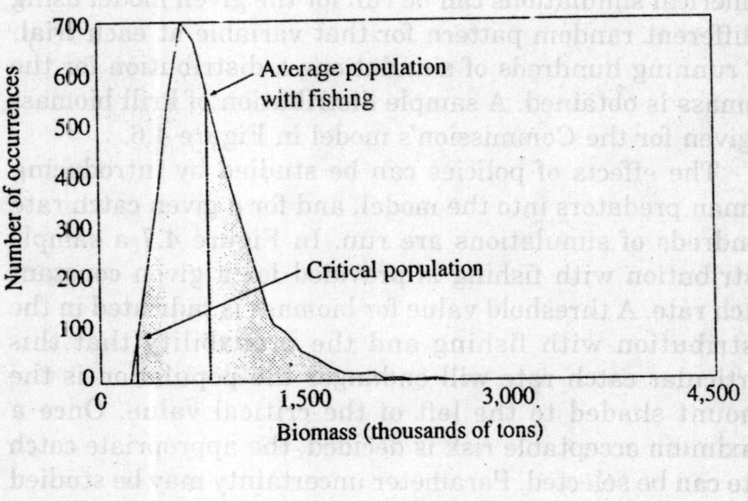

Fig 4.7 : Distribution with fishing.

Monte Carlo integration can be used for which random distributions are assigned to uncertain factors (such as a parameter values or initial condition). Again, hundreds of simulations are run using random values for the relevant factors and a distribution is obtained from these results. A combination of these methods, Monte Carlo for modelling uncertainty and random patterns for natural fluctuations, gives a full range of the direct and indirect effects of the two classes of uncertainty.

Of course, the relationship between Commission action and krill population consequence, the system's feedback mechanism, is not at all clear owing to many factors, but especially to the natural fluctuations of the system. This, in turn, will make it difficult to assess the Commission's performance in managing the fishery. However, as an effort to deal with the 'tremendous uncertainty' of their given ecosystem, the Commission's approach and techniques are both instructive and innovative, and bear watching in the future.

The Law of the Sea

We now turn to a case in which attitudes towards uncertainty about the future helped negotiators reach an agreement. A precipitating event for the Law of the Sea (LOS) Conference was the discovery of manganese modules on deep sea floors. One of the tasks of the LOS Conference was to develop a regime for mining those modules that would be acceptable to the more than 150 countries involved in the negotiations. The agreement was to create a Seabed Authority to collect royalties and taxes on private mining operations and establish an international seabed mining enterprise which would distribute profits among the signing countries. The following description follows from Sebenius (1984), who served as a formal member of the US delegation to the LOS Conference.

The negotiations were interesting from several viewpoints: (*i*) the use of the independent Massachusetts Institute of Technology (MIT) cost simulation model which served to educate negotiators in the intricacies and to provide estimates for the outcome of various regime types; and (*ii*) the sophisticated financial arrangements agreed. The model was deterministic and provided point estimates for variables of interest to the negotiations, in particular, of the internal rate of return and royalties to the Seabed Authority. As negotiations continued delegates became more critical of the basic assumptions and more attuned to the sensitivity of results to such assumptions, for example, the hypothesized discount rate. eventually some of the sources of uncertainty became evident: 'the uncertain capital costs of a new high-technology venture proposed to operate for twenty or more years in a hostile physical environment; uncertain operating costs that depended on energy, chemical, and labour prices; and uncertain revenues that were at the mercy of notoriously volatile metal markets'.

Sebenius maintains that differences in viewing the uncertainties involved actually aided in reaching an agreement. The three principal points of difference are summarized below. On the likely profitability of seabed mining, the G77 (Group of 77) delegates were convinced that enormous profits would be obtained, developed countries projected more modest returns. The negotiated outcome was: low royalty and tax rates on low-success projects, high rates on very successful operations. On risk-aversion, private mining firms would have to invest large sums relative to corporate assets, while revenues from the Seabed Authority paid to each nation would be small relative to national income. Private firms then would be characterized as risk-averse, while individual nations would be risk-neutral. The agreement, which applied higher rates on successful projects, compensating protection of low-end projects, used the diverse risk attitudes to advantage. On time preference, companies evaluate investment projects by determining the after-tax discounted value. The nations involved in the LOS negotiations tended to use lower discount rates as: (*i*) they would receive untaxed revenues; and (*ii*) they were concerned about the long-run viability of the Seabed Authority, and in particular about the welfare to future generations deriving from seabed mining. This difference in time preference was incorporated into the agreement by raising the royalty rate over time so that firms would keep more of the early risk-prone profits and pay higher royalties later which would be heavily discounted anyway.

Much of the success of this novel negotiating approach was attributed to the use of a formal dynamic model. Since the MIT model gave point estimates under the hypothesized model and parameter values, delegates to the conference could test the effects of different assumptions on the estimates obtained for critical variables. For example, the calculated discounted value of the future income stream was seen to vary considerably according to the discount value assumed. Further international negotiation conferences should be able to profit from the experience of the LOS Conference and the agreement's treatment of the inherent uncertainty in environmental economic issues.

Uncertainty and the Control Problem

The optimal control problem briefly reviewed in section 2.3 describes the least difficult conditions for determining an optimal solution path; there are no disturbances and the system model is known. For example, given a particular form for the utility functions, constraints, the maximum pollution level and abatement technology, and given a small enough discount factor, the steady-state solution has saddle-point dynamics. Furthermore, given sufficiency conditions for the normally assumption one can ensure that the steady state is reachable. Finally, given full knowledge on the part of the central authority, the environmental-economic system can be 'controlled' and the economy can be set on the optimal steady-state path.

Sources of uncertainty

Suppose we are seriously considering the possibility of controlling the natural resources of the coupled system through either policy or market instruments or both: what types of uncertainty will enter the control problem? Disturbances may impact the system through various means. Stocks of renewable resources in laboratory conditions, say a controlled fishery, can be predicted with fair accuracy, but there are so many exogenous factor influencing reproduction in open ecosystems such as an open sea fishery, that the authority is encouraged to adopt probabilistic methods and use caution in prediction and policy (see section 3.2). Exhaustible resources are also often difficult to quantify and are subject to random discoveries as well.

Measurement errors will complicate resource control; optimality requires exact knowledge of the current stocks of non-renewables (for the 'Hotelling rule') and renewables (for bioeconomic equilibrium). Moreover, economic and political influences may be a source of uncertainty. In the search for optimal depletion of exhaustible resources such as a copper mine, the owner is confronted with copper prices which are not responding exclusively to the expected total remaining copper stock (if that could be known), but also to other resource prices, economic indicators, geopolitical situations and so on. Of course, policies regarding both renewable and non-renewable resources are subject to government regulation which may be, in part, a function of the state of the resources, but also reflects other exogenous factors. If the uncertainty due to measurement error is represented by a small-variance white noise process, the effects are minor and a process known as 'filtering' out the noise may be used to arrive at fairly accurate estimates of the stocks if the system is linear. If the errors are additive and small-variance white noise, filtering for nonlinear systems may also be efficient; otherwise, measurement errors may be of such magnitude as to compromise the possibilities of optimal resource control.

If the system is disturbed, the consequences are analogous to the effects of observation noise if the system is linear. It is in the nonlinear case that the fundamental difference is obvious. Impreciseness in measurement may mean that we are unable to track the system. But real disturbances may force the system onto a new path, or possibly into a different dynamic behaviour.

In both cases our predictions will be incorrect, but I would argue that it is the second type of uncertainty (from real disturbances that is the most problematic for issues linking the environment and the economic system. However, it may also be argued that due to the stochastic nature of a system such as the climate, it is difficult to convince policy-makers that alarming trends are underway.

Stochastic control

For the control problem, if there is a well-accepted model of the system, and one can characterize the random noise process as additive and small-variance white noise, then the theory of stochastic control can be employed. Consider the problem introduced in section 2.3: the authority can invest national income in abatement activity and also collect emission taxes. There is a well-accepted model for production and utility, but production is subject to random shocks, such as world price levels, which can be modelled as a white noise process. The solution method is that of dynamic programming, which essentially begins with the terminal conditions, maximizes in the last period, and then works backwards, maximizing each step back towards the present. The control problem is to maximize the expected value (the mean) of welfare from t_0 to t_1:

$$\max J = E\left[\int_{t0}^{t1} e^{-rt}\, W(\mathrm{x}, u, t)dt\right],$$

where E is the expectation operator, and this is accomplished analogously to the deterministic problem.

Adaptive control

If the system is subject to observational noise, exogenous disturbances and uncertainty regarding both the system and the combined noise/disturbances, and the system must be controlled nevertheless, the most appropriate approach is that of adaptive control. In adaptive control there are actually two problems which must be solved simultaneously: learning and regulation. On one hand, since the model is not known, it must be estimated. Generally, this means that a linear model is assumed and some algorithm is employed to estimate the parameters. To this end, the system may be designed to probe in order to learn more about the parameters. Basically it disturbs the system and observes the consequences. Suppose the authority is not sure how emission charges will impact emissions. It may decide to try different levels of taxes and observe the reactions. It would be very informative if large changes could be made. On the other hand, the authority also has responsibility for production, and only small perturbations are acceptable here. A relevant implication for questions in environmental economics is that, in the presence of uncertainty, there are trade-offs between learning activities and performing activities, the former favoured by large perturbations and the latter by small ones. Moreover, in these problems, large uncertainty in parameter estimates leads to the introduction of caution learn into the authority's policy. The price of uncertainty will be a further modified golden rule which implies a lower per capita consumption level.

Nonlinear optimal control

Non-constant parameters render the model nonlinear (they multiply variables), and imprecise measurements of the state variables render the problem stochastic. For the general case of the non-linear stochastic optimal control problem, conditions for existence of optimal controls are not known. Even if one assumes that a solution exists and dynamic programming methods are employed, these generally lead to an equation which cannot be solved numerically. One simplification of the problem is to use estimated parameters 'as if' they were the true parameters, an approach known as 'certainty equivalence'. Systems for which linear approximations are not appropriate, such as the general quadratic population model of section 2.1, present all of these problems for determining optimal predictors.

Concluding Remarks

As in the section on dynamics, the most important policy implications to be gained from a study of uncertainty are general. Above all, the policy-maker should come away with an increased awareness of the difficulties of science and formal models to provide the necessary quantitative information on which to base decisions. This wariness will help to offset the over-confidence of some practitioners and to understand why disagreements among experts remain unresolved.

Sources of uncertainty can generally be categorized as to whether they influence the system, such as disturbances or fundamental uncertainty, or whether they only influence our statements (and thus predictions) about the system such as measurement errors, high dimensionality and chaotic dynamics. In either case uncertainty is a limit to our knowledge of the system. This limit must be addressed specifically in the modelling of, and predicting from, systems characterized by, a significant amount of uncertainty.

Uncertainty is represented by random variables and enters the formal mathematical model through noise processes which can be additive and/or multiplicative. Analytical results exist for relatively few stochastic dynamic systems, usually those with few variables and only one or two random variables. The behaviour of larger, integrated models can be studied through numerical simulations of the models. The simulation procedure and mode of presenting results varies with the objective and uncertainty assumed for the exercise.

The krill fishery was used to exemplify the problem of choosing a policy when a great deal of uncertainty surrounds a crucial factor. The juvenile recruitment rate for the krill fishery is subject to natural fluctuations due to environmental factors that cannot be controlled. Then, given a hypothesis for a constant fishing rate, hundreds of simulations are run using different random patterns for recruitment, and the distribution of results can be used as a basis for deciding the correct policy for fishing activity.

Uncertainty can also enter the model due to limited knowledge about a parameter value, such as individual weight gain over krill lifetime. For a small system it may be possible to determine extreme values for parameters so that the worst-and best-case results can be obtained. In larger systems, this procedure will be impossible and a technique called Monte Carlo integration can be applied. Basically, uncertain factors (parameter or initial conditions) are assigned probability distributions, and hundreds of simulations are again used to provide a distribution of possible outcomes. A combination of these two methods will often be necessary.

The policy-maker's concern will depend on the particular case, but he should always insist on seeing the full distribution. A comparison of simple averages may involuntarily omit risks of approaching or surpassing threshold values. One should expect to see increasing use of distributional graphs of numerical simulation results as information input into the decision-making process. This practice will help to educate the policy-maker about the true state of current knowledge, but will also enforce a more complete understanding of the risks and responsibilities involved. The chore of deciding between alternative policies will be rendered more realistic, but also more difficult. Decision-making bodies will have to agree on probabilistic measures which incorporate uncertainty, such as acceptable and probable risk or expected outcome.

In the Law of the Sea negotiations the deterministic MIT model played an essential role. Since the model was formulated as a deterministic model (no uncertainty), predictions were simple point values. However, any change of hypothesis gave a new point value, and delegates to the LOS Conference became adept at using the model to explore uncertainty by considering alternative hypotheses.

Moreover, the LOS agreement actually was able to use disagreement over uncertainty about the future to formulate a sophisticated, contingent financial arrangement accepted by over 150 participating countries.

It is difficult to guarantee optimal controls for the resource-economy problem under uncertainty. Filtering can be applied under restrictive hypotheses (the system is linear and noise is additive and small-variance). In this case, the theory of stochastic control can be applied and dynamic programming can be used to find the optimal control path which maximizes the expected value of welfare.

Adaptive control is a more general method for systems subject to observational noise, disturbances and uncertainty regarding both the noise process and the system. In this approach, the authority must learn about the system (a linear model is generally assumed) and regulate it. The uncertainty regarding parameters leads the authority to include a caution term which would mean a lower per capita consumption for the further modified golden rule.

The general conditions for existence of optimal controls in the nonlinear stochastic optimal control problem are not known. If linear approximation of the nonlinear system does not provide an adequate approximation of the qualitative dynamic behaviour of the system, and there are irreducible uncertainties, optimal control methods are probably not applicable.

Final Note

It would seem that my task has been to emphasize the hard facts of the dismal science of economics. Certainly, confronted with the magnitude of responsibility that the current generation bears for the management of the environment and finite natural resources, one may be overwhelmed to the point of inaction. Yet it is precisely the tricks of dynamics and uncertainty that force the issue, and it is on this point that I hope to have been convincing.

Further Reading

Texts on dynamical systems are generally written with a particular scientific audience in mind, such as engineering students, graduate students in biology and so on, and one should simply choose according to one's background. A non-technical discussion of the real problems created by population dynamics can be found in Hardin (1993). A number of introductions to chaos theory are now available. One such is Devaney (1986), a mathematical text in the spirit of simple discrete models. The literature on dynamical nonlinear systems is vast; Jackson (1989) is a readable reference with useful insights.

To the survey literature cited on the issue of discounting could be added more general material: Price (1993) and Freeman (1993).

Applications of optimal control theory to economics and environmental economics abound. A number of texts which explain the basic procedure are available. I would particularly recommend Chiang (1992) but also, at a slightly more difficult level, Kamien and Schwarts (1991). Well-developed applications to environmental economics can be found in Siebert (1987).

Uncertainty in empirical economics is studied in the field of econometrics, for which there are any number of introductory texts. The treatment of uncertainty in theoretical economics varies according to the particular research programme under consideration. For an example with applications to environmental risk see Bromley and Segerson (1992).

Acknowledgements

I would like to thank Henk Folmer whose suggestions on the text were invaluable and Ignazio Musu who introduced me to the field of environmental economics. Any mistakes and, naturally, all the opinions expressed in this chapter are my own.

Natural Resource Utilization and Environment

Introduction

It is common to divide natural resources into stocks that are capable of regenerating, such as fish populations, and those which are nonrenewable, such as coal or oil. the central feature underlying the distinction between different natural resources is the timescale related to their use. Some resources such as agricultural products, solar radiation and the ability of the environment to absorb non-persistent pollution, may be assumed to be expendable or constant flow resources. The main feature is that use at one particular point in time will not affect the amount that can be used in the future (see Table 5.1).

Salt is included in expendable resources because of its large amount and its easy availability; it is difficult to imagine that there will ever be a shortage. In contrast with using expendable resources, using renewable resources such as fish populations and forests usually affects future utilization possibilities. This means that rational utilization policy must take into account the time dimension during which the resource is regenerating. Biologically regenerating resources are often characterized by emphasizing that they are 'renewable but exhaustible'. This refers to the minimum viable population size. Harvesting a population.

Table 5.1: Example of natural resources.

Available	*Biological*	*Physical properties*		
		Non-energy minerals	*Energy*	*Environmental*
Expendable	Most agricultural products, for example, Corn, grains	Salt	Solar radiation Hydropower Ethanol	Noise pollution Non-persistent air and water pollution
Renewable	Forest products Fish Livestock Harvested wild animals Wood Whales Flowers Insects		Wood for burning Hydropower Geothermal power	Groundwater Air Persistent air and water populations animals populations forests
Depletable	Endangered species Peat*	Most minerals, for example, gold, Iron ore, bauxite Top soil	Petroleum Natural gas Coal Uranium Oil shale	Virgin wilderness Ozone layer Water in some aquifers

smaller in size may cause extinction. The other threat to these resources is the deterioration of the environment in which they are regenerating.

In the case of nonrenewable or deplorable resources, consuming a unit of the resource implies that the stock for future consumption is reduced forever. However, most depletable resources, such as oil and coal, are in fact regenerating but, from the human time horizon point of view, at a regeneration rate of approximately zero. It is evident that when using these kinds of resources the main question is how the benefits from the resources should be allocated rationally and equitably to different generations.

Inclusion of the timescale implies that natural resources are considered capital assets in economic terms. Allowing a renewable resource stock to grow is investing in future harvesting possibilities; decreasing the stock size is disinvesting. Rational decision-making requires the comparison of the costs and benefits accruing at different moments of time.

Natural resource utilization is closely related to pollution problems. This is due to 'materials' balance, which means that human production activity ultimately leads to depositing the utilized material back into the environment as 'pollution'. One implication is that efficient pollution control is impossible if the rate of natural resource utilization is too high. An example of this is the former Soviet Union, where low prices of natural resources made recycling unprofitable and did not provide any incentives for emission abatement. Another close connection between natural resources and pollution problems is that the economic analysis of both problems requires a long timescale, that is, the welfare of future generations must be included in the present environmental policy. How this can be done is one of the most complicated but most studied questions in natural resource economics. Understanding how natural resources can be allocated efficiently to different generations gives direct insight into long-term pollution problems as well.

Economists have been studying the scarcity of natural resources since David Ricardo (1972-1823) and John Stuart Mill (1806-73), with an abundance of literature as a result. To present the most essential principles we have chosen three typical examples of natural resources. We will first study nonrenewable resources such as minerals and fossil fuels. Secondly, we will consider economics of fisheries and the problem of open access. Finally, we will discuss forest economics. However, before going on it is necessary to have a close look at the role of property rights in natural resource utilization. As we will see, the difficulty of assigning property rights to nature is the key economic explanation for too fast a depletion of natural resources.

Property Rights and Natural Resources

Property rights can be understood as characteristics that define the rights and duties for using a particular asset or resource. According to Bromley (1991), natural resources can be utilized under some of the following property rights regimes:

1. Under a state *property* regime the ownership and utilization of the resource is being controlled by the state. Individuals may be allowed to use the resource, but only according to the rules imposed by the state. Examples are state-owned national forests, parks and mines.
2. Under a *private property* regime, the right to utilize the resource, as well as buying and selling it, is controlled by individuals. Examples are private forests and meadows.
3. *Common property* regime means that a group of owners can control the use of the resource and prevent others from using the resource. The members of the group have specified rights and duties. Examples are common land, in some cases the harvesting of animal populations in ancient societies and, perhaps. 'every man rights' in Nordic countries.

4. Under an *open or free access* regime, each potential user of the resource has complete autonomy to utilize the resource since no one has the legal right to keep any potential user out. One cannot speak of property since there are no property rights. An example is whale populations when countries had not yet signed international agreements for regulating excessive harvesting.

It is obvious why property right, or the absence of them, have played a central role in natural resource economics. A particular natural resource stock creates economic surplus, which is the difference between the market price of the resource and the costs incurred when the resource has been extracted or harvested. This difference represents the value of a particular natural resource unit and is called 'economic rent'. When this rent is received by a particular resource owner, and when he can be sure that the situation will continue in the future (as in the case of private or common property regimes), it is in the interest of utilizers to manage and harvest the resource carefully. Under the free access regime the situation is completely different. There is no resource owner and those who utilize the resource have no incentives for careful management or conservation of the natural asset. What is absent is a binding agreement which assures every user that if they stop the overexploitation the other users will behave similarly.

Biologically renewable resources are often examples of free access resources. In 18th-century Sweden, wood-based industries had free access to forest resources (Johansson and Löfgren, 1985). As a consequence, the condition of forests was worse than at present. North Sea herring was harvested in a free access situation in the 1970s until fishing was temporarily prohibited by an international agreement (Björndal and Conrad, 1987). American bison was exploited to near extinction is a free access situation. At present the reindeer pastures in Finnish Lapland are free access or badly managed common property resources, although the reindeer are privately owned.

This chapter emphasizes that property rights are important when explaining the bahaviour of a natural resource industry and resource markets, but also from the point of view of resource policy. In a market economy the behaviour of firms is characterized by a profit incentive which, in the case of natural resources, means the desire to maximize the economic rent from resource utilization. The definition of property rights may completely determine the outcome of profit maximization. We will see, especially in the section regarding fishery resources, that efficient and equitable intervention of public authorities easily fails when the interplay of ill-defined property rights and the profit incentive is neglected.

Nonrenewable Resources

The economic definition of nonrenewable or exhaustible resources state that the intertemporal sum of the services provided by a given stock of an exhaustible resource is finite (Dasgupta and Heal, 1979). This definition highlights the importance of the long-term perspective. As we will see in this chapter, the central economic problem of these resources is how to allocate their fixed amount intertemporally, that is, between different points over time or between different generations.

It is difficult ot overemphasize the importance of nonrenewable resources for industrialization and the modern economy. Energy production in developed countries is heavily based on oil, gas, coal and uranium. Similarly, it is difficult to imagine present industrial production without aluminium, lead, plastic or iron, to mention a few examples. As a consequence, there has been serious concern as to if and when the limited availability of these resources will set an upper bound to economic growth and welfare. Another frequent debate is whether the market economy, with economic growth as one of its main objectives, will exhaust these resources in a myopic manner, that is, without taking into account the welfare of future generations. In other words, is the rate of extraction by a profit-maximizing mining industry appropriate from a social point of view?

The discussion about the above-mentioned issues has often been dominated by gloomy predictions. Let us mention some examples. In 19th century England it was predicated that there would be enough coal for heating for 30 years. In 1874 a US geologist predicted that petroleum reserves would last four years. In 1934 a group of geologists predicted that world copper reserves would be depleted in 40 years. In 1974 the same prediction method indicated, however, that there was still enough copper left for 57 years. The famous 'Limits to Growth' report of the Club of Rome, published in 1972, predicted that gold, mercury, silver, tin and zinc would all be exhausted in the 1980s.

One reason for the fortunate failure of these predictions if that the scarcity of nonrenewable resources is viewed purely in physical terms. This perspective has led to the use of a static reserve index (or reserve-to-use ratio) which measures the ratio of current reserves to current consumption. Current reserves are the known resources that are economically extractable. The index expresses the number of years until the resource is depleted, given that there will be no additions to the reserves and that the future annual use of the reserve remains at the current level. Both of these assumptions seem to be seriously flawed. First, because of discoveries of new deposits the size of reserves may increase over time despite continuing extraction. In addition, the size of reserves increases due to rapid technological development, which makes the extraction and utilization of known subeconomic stocks less expensive. Secondly, the static reserve index neglects technological development in recycling and the possibility of substituting scarce materials for more abundant resources. From this follows, as in the case of oil and gas, that reserves have increased over time.

Let us consider the economic principles underlying the analysis of non-renewable resource scarcity.

Nonrenewable Resource Use According to the Model by Hotelling

In a classic article published in 1931, Hotelling defined the question of non-renewable resource use as that of allocating a given amount of resource stock over different moments of time in order to maximize the utility or benefit from consuming the resource. During the past 60 years the economic analysis of this question has become very complicated compared with the one carried out by Hotelling. However, it is possible to say that the basic insight and results of Hotelling are still included in nearly all economic analyses of nonrenewable resources. The analysis of Hotelling was carried out using complicated mathematics which deterred many economists of his time from understanding his ideas, although the basic economic logic of the analysis was intuitive.

The first notion in the Hotelling analysis is that the size of a nonrenewable resource stock is fixed. Thus consuming and extracting a unit of the resource now implies that there is less of the resource in the future. This means that, in addition to extraction costs, there is another special kind of cost: the reduced level of future benefits due to fewer resources being available. In the context of nonrenewable resources the amount of reduced benefits is called 'royalty' or 'user costs'. These costs reflect the scarcity of the resource and must be taken into account when deciding the extraction policy. Hotelling assumed, as is usual in theoretical economic analysis, that the marginal utility of consuming the resource is decreasing. In addition, he assumed the marginal extraction costs to be constant. This means that along an economically optimal extraction path the marginal utility of consuming the resource must be equal to the sum of constant unit extraction costs and the royalty. Thus marginal utility exceeds the marginal cost of extracting the resource. The difference, that is, royalty, reflects the value of an unextracted marginal resource unit with respect to future consumption possibilities. Royalty means that society must be more conservative in consuming nonrenewable resources compared with ordinary goods, the production costs of which do not include royalty.

The idea of royalty and optimal utilization decision is explained in Figure 5.1. the vertical axis measures marginal utility from consuming the resource, marginal costs of extracting the resource and the price of the resource. The horizontal axis represents the level of resource utilization over one year. As is shown, the marginal utility of consuming the resource is declining. With normal commodities, the optimal level of producing the commodity is reached so that marginal utility equals marginal production costs. This condition is reached at level q**. However, in the case of nonrenewable resources, consuming the resource today implies opportunity costs, that is, the royalty. If the level of royalty equals A–B, then the optimal level of resource extraction and consumption is q*.

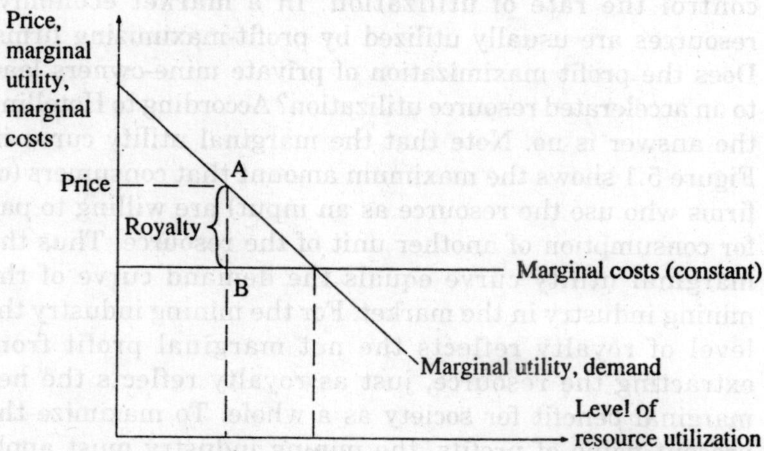

Fig. 5.1 : Royalty and the optimal utilization decision.

The next question is how the level of royalty evolves over time. hotelling derived the result that royalty must increase at a rate equal to the rate of discount ('the Hotelling rule'). This means that the net marginal utility of consuming the resource must also increase at a rate equal to the rate of discount. The interpretation of this result is an intuitive one: it implies that the discounted value of net utility is the same in all periods. In other words, from the current point of view the net benefits from the marginal units consumed in every period are equalized. This equality implies that it is impossible to increase the total level of discounted net utility from resource consumption via changing the extraction of the resource over time. The Hotelling rule implies that royalty increases exponentially (an increase of for example, 5 per cent every year) and the level of resource utilization decreases over time. the higher the rate of discount, the higher the rate at which the level of resource consumption falls over time.

So far our discussion has described optimal nonrenewable resource utilization when society can directly control the rate of utilization. In a market economy, resources are usually utilized by profit-maximizing firms. Does the profit maximization of private mine-owners lead to an accelerated resource utilization? According to Hotelling the answer is no. Note that the marginal utility curve in Figure 5.1 shows the maximum amount that consumers (or firms who use the resource as an input) are willing to pay for consumption of another unit of the resource. Thus the marginal utility curve equals the demand curve of the mining industry in the market. For the mining industry the level of royalty reflects the net marginal profit from extracting the resource, just as royalty reflects the net marginal benefit for society as a whole. To maximize the present value of profits, the mining industry must apply such a rate of extraction that royalty increases exponentially over time. but this means that a resource extraction policy that maximizes the profits for the mining industry equals the socially optimal resource utilization policy. In addition, as Figure 5.1 indicates, the higher the level of royalty, the higher the market price of the resource, which implies that the market price of a nonrenewable resource must increase over time.

Extensions to the Hotelling Approach

Following the work by Hotelling, several additional features have been incorporated in the above analysis. The present models take primarily into account technological development in resource extraction and the discovery of new deposits. Both of these imply that the market price of nonrenewable resources may have a U-shaped form (see Slade, 1982), that is, in an historical perspective the market price may first fall and later start to increase. This occurs because technological development and new deposits will dominate the increase in royalty at the beginning or resource extraction. Later the increases in royalty and increases in extraction costs will dominate, causing the prices.

That resource extraction by profit-maximizing firms equals the socially optimal policy seems to hold only under restrictive assumptions. There are at least two situations in which profit-maximizing firms may extract resources too rapidly. First, private firms may discount future profits by using a rate of discount that is higher than the preferred one from a social point of view. Secondly, extraction of nonrenewable resources usually causes environmental damage, that is, external effects. If these problems are taken into account, then the rate of optimal resource utilization will be lower than that in a private mining industry. How can society influence the rate of resource extraction? One possibility is to apply a tax per unit of resource extracted. This kind of tax, if accurately implemented (Sweeney, 1993), may shift the resource extraction from the present to the future and thus promote better conservation of resources.

Have Nonrenewable Resources Became More Scarce?

We started this discussion of nonrenewable resources by referring to predictions based on the static reserve index. As could be seen, this measure seems to give a seriously biased picture of the scarcity of nonrenewable resources. A completely different view is given in an influential study by Barnett and Morse (1963). According to their work, the relative prices of minerals and nonrenewable resources have, in general, declined in the 20th century. This result has been widely interpreted as suggesting that there is no reason to worry about the increasing scarcity of nonrenewable resources, thanks to technological development, substitutes and new discoveries. However, more recent evidence (Fisher, 1981; Slade, 1982) suggests that, regarding several minerals, a turning point has been reached which means that their prices compared with other resources and commodities have begun to rise. However, most economists in this area argue that, because of numerous substitution possibilities, the limits of nonrenewable resources will be found in the form of increasing extraction costs rather than in the form of the physical limits of the resources.

As carly as 1848 John Stuart Mill argued that a serious problem of nonrenewable resource utilization may be the environmental side effects of extraction activities. This view is now shared by many economists because of, for example, the accumulation of carbon dioxide in the atmosphere.

Economics of Fisheries

The economics of fisheries includes a particularly vast selection of various problems in natural resource utilization. Fish stocks are populations with complex biological reproductive dynamics. The growth and productivity features of these natural assets must be known and incorporated into economic analysis before rational utilization decisions or the behaviour of the fishing industry can be understood. This means that the economic analysis of fisheries should be interdisciplinary, using models which combine biological and economic facts.

One important result of regarding fish populations as natural assets is that a long-term perspective is inherent in the economic analysis of fisheries. In particular, the literature contains and interesting debate between the maximum sustainable harvest of fish populations and the economically efficient utilization policy. The economics of fisheries also shows how aspects other than the raw material value of natural resources can be incorporated into rational utilization policy.

The major issue in this area follows from the gap between the nature of institutions and the nature of fish populations. These populations are usually very mobile, and it has been difficult to assign property rights to individuals or even nations. As a consequence, fish stocks are harvested in many places in the world in situations with strong open access features. One main theme in the economics of fisheries has been the analysis and development of public regulation measures that guarantee sustainable harvesting policies and positive net revenue for the fishing industry.

The Debate Between Maximum Sustainable Yield and Economic Optimality

Fish populations can be harvested indefinitely if the size of the population of the environment in which they are reproducing is not in danger. However, a given fish population can be harvested on a sustainable basis by applying various different harvesting intensities. This can be demonstrated by a very simple growth model shown in Figure 5.2. the diagram depicts the growth of a fish population over a year as a function of the size of the population, measured as the weight of its biomass. According to this biological model, the growth of the population is low when the size of its biomass is low and when the size of its biomass is near the carrying capacity level, represented by symbol K. if there is no harvesting, then the population grows until it reaches its carrying capacity level. The carrying capacity is the maximum level of population that the given habitat can support.

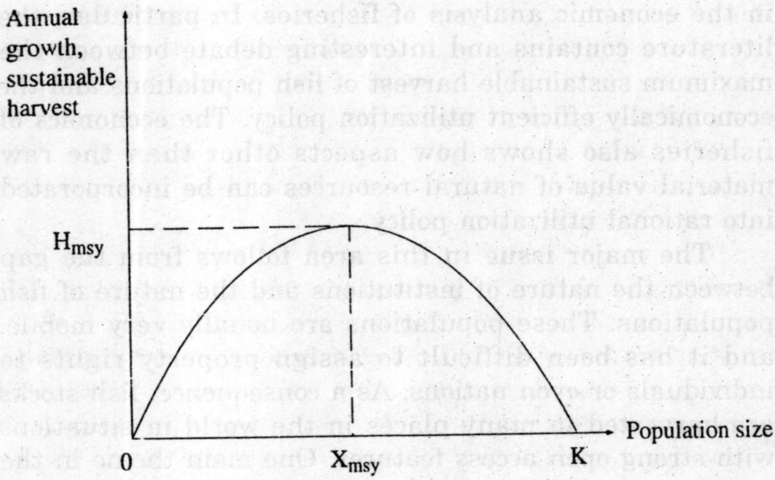

Fig. 5.2 : Maximum sustainable yield

Perhaps the most traditional approach in the utilization of fish resources, often promoted by fishery biologists, is to apply the maximum sustainable harvesting policy, that is, the 'full utilization' of the resource. In terms of Figure 5.2, this is a policy which maintains the stock at the level represented by X_{msy}. This means that the annual harvest of the resource can be kept at the level H_{msy}. Note that any harvesting level above H_{msy} will cause the population to become extinct. This kind of utilization policy seems to be the most preferable policy because it gives the largest possible material flow that is sustainable over time.

However, there are at least three reasons why the policy of maximum sustainable yield is not optimal from an economic point of view. First, because the fish population is regarded as a capital asset, all units of the population must yield a positive rate of interest just as other assets in the economy. To yield interest, and asset must 'grow'. If the fish population is kept at level X_{msy}, then the marginal or last unit of the population does not grow. Note this from Figure 5.2: if the population size equals X_{msy} and we decrease the size a small amount, the loss in the total annual growth of the population is approximately zero. This is because the growth curve is flat when the population equals X_{msy}. Thus the last unit of the population yields a zero rate of interest and the size of the population is too large from an economic point of view. In contrast, if the amount of money received from selling the additional harvest of one year is invested in some other way (for example, in a bank account) a positive rate of interest will be obtained. This means that it is profitable to decrease the level of population from X_{msy} as long as the contribution of the marginal population unit is lower than the appropriate interest received from selling the catch and investing the money in the most profitable way. Thus the fact that the owner of a capital asset can receive a rate of interest in a market economy implies that, for the owner of a fish population, it is profitable to maintain to stock below the level of X_{msy}.

However, there are two other things to be taken into account. The cost of harvesting the fish population tends to rise as the stock size is reduced. The reason is that the smaller the population, the fewer fish the fishermen can harvest given fixed working hours and fishing capacity. This implies a positive effect on the optimal size of the population. Another reason for keeping up a larger population size is that the stock may be valuable *in situ*. For example, in the case of sea mammals many people share the view that these animals have existence value (or, for example, observational and photographic value) and that the correct size of the population cannot be determined by taking into account only the value of the raw material received. This implies a positive effect on the economically optimal population size.

The requirement that a fish population as a capital asset must yield a rate of interest implies that the stock size may be smaller than X_{msy}, while the dependence of harvesting costs on the stock size and the *'in situ'* value of the animals implies a larger stock size. The economically optimal population size and harvesting level depends in each case on the relative strengths of these different effects.

It is important to note that if the property rights of a fish population are well defined (for example, that there is one owner of the population), then profit maximization is usually in line with harvesting the resource on a sustainable basis. However it is important to ask whether economic rationality and carefully defined property rights guarantee that the stock is not driven to extinction. This important question was first studied by Clark (1973).

Let us assume that the cost of harvesting the resource is independent of the size of the stock and that only the raw material value of the population matters. If the rate of interest in the economy is high compared with the growth rate of marginal population units, then the owner of the population may always think it profitable to decrease the size of the population and invest the returns in a more profitable way in the capital market. This shows that, especially if *in situ* or existence value is not taken into account, there is nothing which *a priori* guarantees the optimality of sustainable harvesting policy. However, in reality whether it is profitable to harvest the entire stock is the exception rather than the rule, because the harvesting costs in the case of small population size may be too high compared with the market price of the resource. This usually guarantees the optimality of harvesting the stock at a sustainable basis. It is likely that renewable populations are only overexploited in an open access harvesting situation.

Dynamics in Open Access Fisheries

Established fisheries in different parts of the world are often characterized by a situation in which the profit of the fishing industry is low and the level of harvest is only a small proportion of its level in the past. The economic model of open access fisheries explains the dynamic economic forces that can create this kind of resource waste. It is useful to consider the underlying economic forces of open access resource utilization because without this understanding it is difficult to design a regulation policy which effectively and equitably corrects the problem of overexploitation.

The cost of harvesting a fish population consists of two distinct components. First, there are wage expenses and the costs of fishing gear, fuel and so on. These are the private costs for each fisherman. The other cost component cannot be directly observed. It is an implication of the limited size and limited reproductive capacity of a fish stock. Because of the limited size of the fish stock additional harvesting always means lower harvesting possibilities in the future. In other words, an uncaught fish left in the sea is valuable because it guarantees the future harvest and profits for the fishermen. The value of uncaught fish depends on many different factors, such as the rate of interest, future markets for fish, technological development of fishing gear and so on. If the fish population has a private owner whose harvesting policy is in line with the principles descibed in the section above, then he will approximate 'the value of the fish in the sea' by using all his knowledge of fish markets, cost of fishing, the rate of reproduction of the stock and so on to maximize his stream of profits. The value of uncaught fish prevents a rational fisherman from overfishing his stock. In this way, rational profit maximization usually guarantees a conservative harvesting policy. In contrast, when the fish stock is harvested under an open access regime, no fisherman can be sure of who will benefit from the value of an uncaught fish. Thus fishermen have no incentive whatsoever to take the price of the fish in the sea into account when they make decisions about their individual harvesting levels. In a free access situation, no individual fisherman can efficiently conserve the resource by refraining from harvesting. Doing so will only enhance the harvesting opportunities for his competitors, which is the tragedy of free access harvesting.

According to the model of Smith (1968), evolution of a free access fishery may typically proceed as follows: at the initial stage, the fishing industry may earn very high profits because the industry is exploiting the natural resource at an excessively high rate by neglecting the value of the unharvested resource. It is thus earning present and future profits at the same time. In addition, the industry's profits include resource rent, which belongs to the owner of the resource as a compensation for having a valuable capital asset. However, because of free access there is no owner of the resource. This means that the short-term profit level in this industry exceeds the normal profit level in the economy. Because there are no property rights, the high profit level for the fishermen makes the fishing industry attractive to new fishermen, who will enter the fishing industry. As a consequence, supply increases and the price of fish in the market will fall. This will reduce earnings in the fishing industry to the level where the profit for the fishermen is nearly zero. In addition, the higher level of fishing means that the size of the fish population may decrease, which means that the profit of a typical fisherman will eventually become negative. Some of the fishermen may be forced to leave the business and sell their boats at a low price. If private harvesting costs are insensitive to the size of the fish population, extinction may occur. In many cases the costs of overfishing a small fish population are formidable. This prevents the industry from destroying small its own resource basis by Gordon (1954), the evolution of the fishery may stabilize to a situation called 'bionomic equilibrium'. At this equilibrium, the population size, harvesting levels and the number of fishermen are approximately stable. The situation is, however, economically inefficient. The fishing industry is overcapitalized. In other words, there is an excessive number of fishing vessels compared with the low levels of harvest and fish population. As a consequence, the whole economic rent, that is, the amount of net economic benefit which can be derived from the renewable resource, is wasted. Compared to this situation it is theoretically possible to have fewer vessels harvesting a larger fish stock and thus yielding a positive level of profits, a portion of which may be used to compensate those fishermen who were asked to leave the industry.

Perhaps the most serious open access situation is when different countries are exploiting the same fish stocks. An interesting example of the dynamics of a free access fishery is described in a study by Björndal and Conrad (1987). They analysed the open access exploitation of North Sea herring during the period 1963-77. The stock was harvested by Norway, Denmark, the Netherlands, Germany and the United Kingdom. The evolution of the fishery closely resembles the theoretical model described above. Figure 5.3 shows that in 1963 the size of the fish population was about 2.4 million tonnes, and that there were about 120 fishing vessels. Over the next ten years the size of the population declined to one million, but the number of fishing vessels increased six-fold. The low level of the population caused harvesting costs to increase and after 1972 the number fishing vessels started to decline. In 1977 the authorities realized that the stock was seriously overharvested and Norway and the European Union agreed to cease fishing. The model of björndal and Conrad predicted that, without banning, extinction might have occurred in 1983. Under the moratorium, which lasted until 1981, the stock recovered and it was estimated that in 1983 the stock level was 0.6 million tonnes.

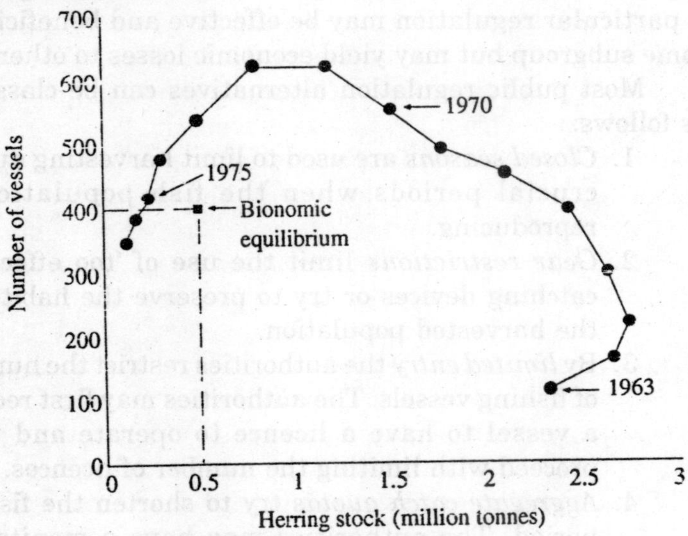

Source: Björndal and Conrad, 1987.

Fig. 5.3 : Dynamics of an open access factory of North Sea herring.

It is important to realize that in an open access situation there are no 'automatic' economic forces that can solve the problem of overexploitation. Fishery economists agree that public intervention is necessary to achieve economic efficiency. However, there are numerous examples of public authorities having made the situation even worse because of taking inappropriate regulatory measures. Let us consider these problems.

Regulation of the Fishery

Public authorities have taken various control measures to reduce the problem of overcapitalization and to save population from extinction. At practical level, regulation of fisheries is in many cases an unsolved problem (Munro and Scott, 1985). The authorities seem to neglect the economic forces behind overexploitation. This has led to regulation measures which may temporarily reduce the level of over-harvesting, but which still include the market signals that cause overexploitation. The other problem is that fishery regulation faces the conflicting interests of different groups. A particular regulation may be effective and beneficial to some subgroup but may yield economic losses to others.

Most public regulation alternatives can be classified as follows:

1. *Closed seasons* are used to limit harvesting during crucial periods when the fish population is reproducing.
2. *Gear restrictions* limit the use of 'too effective' catching devices or try to preserve the habitat of the harvested population.
3. By *limited entry* the authorities restrict the number of fishing vessels. The authorities may first require a vessel to have a licence to operate and then proceed with limiting the number of licences.
4. *Aggregate catch quotas* try to shorten the fishing period. The authorities may have a monitoring system for ceasing fishing when the cumulative harvest level equals the aggregate quota.

5. *Taxes* can be imposed on the catch or on some harvesting input.
6. *Individual transferable quotas* limit the level of harvest for each individual fisherman per fishing period.

The aim of closed seasons and gear restrictions is to reduce the effectiveness of harvesting or increase the real cost of fishing. The reason for these policy measures seems to be a biological one. The aim is to reduce the level of harvest to the maximum sustainable yield or below, in order to protect the stock from over harvesting. Even if these goals are met they may still cause economic inefficiency and waste of resources, that is, excessive working hours for fishermen and large expenditures on capital and fishing equipment to achieve the desired level of harvest. Economic efficiency requires the level of the annual harvest to reflect the value of the fish in the sea and to harvest at the lowest possible cost. By increasing the cost of harvesting, the regulation policy reduces the income level of fishermen and increases the price of fish for the consumer.

Limiting the number of vessels in a particular fishing area may first reduce the level of harvest, but in the long run this policy will easily lead to a phenomenon called 'capital stuffing'. This means that, while the number of vessels is restricted, the harvesting capacity of existing vessel is increased by increasing the size of engines, using larger nets or by adding more electronic devices to find the schools of fish more efficiently. Thus this form of regulation causes economic losses without the guarantee that the problem of overharvesting is solved.

The Baltic Sea salmon fishery serves as an example in which the Finnish authorities monitor the cumulative harvest during the fishing season and ban fishing when the aggregate catch quota is reached. The implications of this kind of regulation are quite straightforward. The regulation may increase the economic inefficiency of free access harvesting. The fishing period starts in early summer when the catch will remain fresh for only a few days. However, fishermen have a strong incentive to start harvesting at full capacity because being too slow at the beginning of the season may imply that their annual share of the total quota will be small. Fishermen must invest in strong engines and use their time as effectively as possible to be able to compete successfully with other vessels. This implies that there is an oversupply of salmon during the summer and that the price of fish is at its minimum. The season is closed during the automn. At Christmas the demand for salmon is at its maximum, but there is no supply and salmon are imported from Norway. In general there is no guarantee that this kind of regulation will increase the welfare of the fishermen or consumers (Mickwitz and Pruuki, 1992).

On the face of it a tax on catches or landings appears to be very appealing. The fundamental problem of the open access fishery is that fishermen have no incentive to take into account the value of fish left in the sea. Public authorities can impose a tax on a unit of catch, which will represent these neglected costs. The aim of profit maximization forces the fishermen to reduce the level of harvest and, if the authorities have calculated the tax level accurately, and optimal harvesting policy will follow. The problem is, however, that this regulation measure may be politically unfeasible. The reason is simply that the tax transfers the economic rent from the fishing industry to the public authority. In other words, the tax policy may imply an economically optimal harvesting policy but it redistributes income between the government and the fishing industry. As a consequence, fishermen will use all their political strength to prevent this kind of policy from coming into force (Munro and Scott, 1985).

Under the system of individual transferable quotas (ITQ) the authorities determine the optimal seasonal aggregate catch and subdivide it among the fishermen. In this way the government links property rights to the harvested population. The individual quotas allow the holder to harvest a given amount of a specified type of fish.

In addition, it is possible to sell the quotas to other fishermen. Consequently, the fishermen have no incentive to overcapitalize. They can freely decide on the fishing equipment they want to use and on the length of the fishing season. Moreover, it is possible to sell part of the individual quotas in a particular year or to leave the business completely. The ITQ system has the same effect as a tax on catches, but it leaves the net economic welfare (that is, the economic rent) to the fishing industry. The main difficulty of this system is the initial allocation cf quotas of the vessels. One possibility is that the authorities give quotas in proportion to historical catches. The ITQ system has been applied in New Zealand from 1983 onwards and to some extent also in the USA and Canada.

Economic analysis of fisheries provides an interesting perspective on the utilization of natural resources. In many respects this analysis may also be extended to other biological resources such as game. Natural resource economists share the view that renewable stocks should be considered capital assets with a biological reproduction mechanism. Special attention must be paid to the definition of property rights. Without combining economic and biological perspectives, and without appropriate property rights or correctly designed regulation policy, the economic benefits from these resources, including the value *in situ*, is easily wasted.

Economics of Forestry

Forests cover 35 per cent of the total land area in Europe whereas the world average is 32 per cent. They provide a large variety of materials and services to world economies. These include raw material for housing, wood products, and pulp and paper products. However, this is by no means all; forest economists have tended to overlook services other than wood production when considering the optimal management of forests. This may be partly due to the common property or even free access nature of most of the other services. However, the increasing importance of environmental problems has also directed attention to the non-timber services. Forests provide habitats for wildlife, recreation possibilities for humans, absorb carbon dioxide and produce oxygen. Forests also help to preserve water resources and prevent erosion.

The traditional problem for forest economics has been the optimal management of forests with respect to timber production. In view of what has been said above, this is clearly a very narrow perspective. Instead, society should find an optimal mix of all the different services provided by forests. It is obvious that in different regions the optimal use of forest resources can vary considerably. In some cases, for example, in managed second growth boreal forests, even intensive use for timber production may not seriously conflict with the other uses or distort the ecological balance of the environment. In contrast, some forests should be used very little, if at all, for timber production. For example, the boreal forests near the forest borders in the northern parts of the European, Asian and American continents, and the rain forests where the ecosystems are especially vulnerable, should be managed with great caution.

The present situation indicates an unfortunate polarized development in the utilization of the world's forest resources. In Europe, where timber production mainly occurs in managed and plantation forests, the resource is increasing. The same holds for most managed boreal forests, although in the United States and Canada, felling old-growth timber is still a part of the raw material base of the wood processing industry. In contrast, in the tropical regions deforestation is continuing at an alarming rate (World Resources Institute, 1990). Deforestation results not only in reducing wood resources, but also in intensifying global warming and erosion, and decreasing agricultural production and biodiversity.

There is no easy solution to the above-mentioned problems, and in this section we can only give a short description of the 'traditional' problem of forest economics which answers the question of when to harvest a stand of trees. After a description of tree growth, we define the 'mean annual increment' (MAI) or the mean biological yield of a stand and introduce two different criteria to determine the optimal harvest time. this is followed by a discussion of the most important applications and extensions, and by a brief description of the expected development of European forestry resources. For a more detailed discussion of other aspects of forest management and further reading, one is referred to, for example, Tietenberg (1992).

Approaches to Optimal Rotation Period

in 1849, more than 100 years before the dynamic problem of fisheries management was presented, the dynamic optimization problem in forestry—the optimal rotation problem—was posed by a German forester, Martin Faustmann. It was solved 11 years later by another German, Max Robert Pressler (Johansson and Löfgren, 1985), which resulted in the optimal moment for harvesting an even-aged stand.

Practical forestry was, and still is, often based on criteria other than the economic principle provided by the faustmann solution. Samuelson (1976) showed that from an economic point of view the correct way of determining optimal harvest time is still the one provided by Faustmann, although other criteria have been proposed. This renewed theoretical interest in the optimal rotation problem, and several authors have since presented generalized and extended result (for example, Hartman, 1976; Heaps and Neher, 1979; Newman *et. al.*, 1985; Mitra and Wan, 1985, 1986), but these are beyond the scope of this book.

The rapid development of dynamic programming and optimization methods since the 1950s, and of efficient computer software since the 1970s, has made the applications of (economic) optimization accessible to practical forest management. Therefore, it has also become easier to take the other services of forests into account when management decisions are to be made.

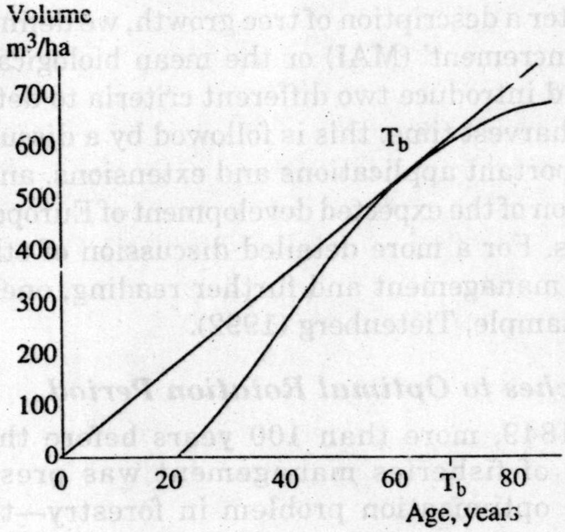

Fig. 5.4 (a) : Volume growth (yield) in an even-aged pine stand. Volume of the trees as a function of time (age of trees in years) in an even-aged planted Norway spruce stand in Finland.

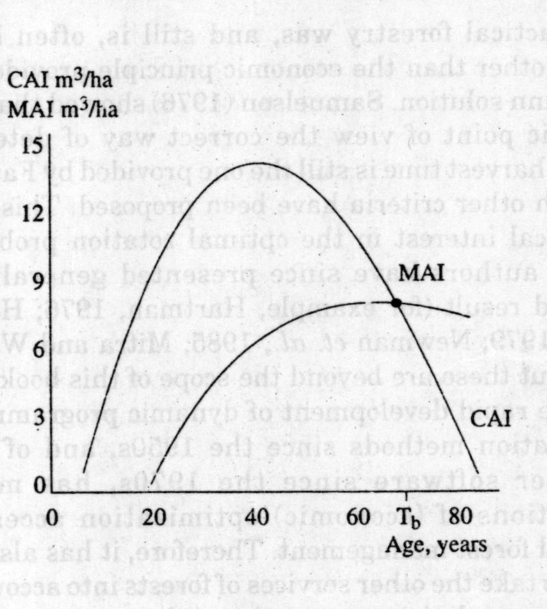

Fig. 5.4 (b) : Maximization of means annual increment.

In the basic optimal rotation madel the volume of an even-aged stand (or a tree) is assumed to depend only on its age. Volume development of a Norway spruce stand in Finland is depicted in Figure 5.4 (a) (Valsta, 1992:7). The horizontal axis represents and age of the stand in years and the vertical axis represents the tree volume in m^3/ha. For very young trees, volume growth is slow. This is followed by a phase of a rapid volume increase. Finally, as the stand matures the relative growth shows down until it stops or is reversed due to natural mortality. In reality, the growth of a tree in a stand also depends on the tree species, its size, the quality of the soil, the density of the stand and so on. These factors are, at different degrees of sophistication, accounted for by the practical applications to be discussed later.

Foresters often determine rotation length from a purely biological point of view. Forest policies can be implicitly or explicitly aimed at maximizing the physical timber production. At the stand level this means maximizing the MAI of the forest stand (compare with the maximum sustained yield in the fishery model). This is calculated by dividing the accumulated timber volume by the age of the stand. MAI for the stand in Figure 5.4 (*a*) is depicted in Figure 5.4 (*b*). it reaches its highest value at time T_b. if trees are harvested each time when the stand reaches this age, then a sustained biological yield of the forest site is maximized over a perpetual chain of rotations, from Figure 5.4(*b*) we can conclude that the current annual increment (CAI) (that is, the increment measured over a given time interval, here over a year) is higher than MAI before time T_b and lower after T_b. One way of finding the rotation length that maximizes the physical timber production is thus to find the age of the stand when the MAI is equal to the CAI of the stand. The other graphical way to find this point is to simply draw a straight in Figure 5.4(*a*) from the origin so that is just touches the 'volume per hectare' curve.

Economically Optimal Rotation Period

Maximizing the MAI is not an economic solution. It does not consider forest a capital asset. Thus, T_b is generally not the economically optimal rotation time. Maximizing the raw material flow from the forest site need not be the same as maximizing the present value of the net economic benefits of the forest site, which is the goal from the economic point of view. The problem formulation to find maximum MAI actually contained no economic parameters. In order to find out the economically optimal rotation period one needs to calculate the present value of the net revenue of planting and harvesting the stand. Thus one crucial parameter of the problem is the decision-maker's rate of discount, represented for example by the long-term interest rate in the capital market.

Assume that the cost of planting is given, say 5,000 ECU/ha, and the timber prices known and will remain constant in the future at, say, 150 ECU/m^3. We assume that harvesting does not cost anything. Now the question is at what age to harvest the stand, the development of which is depicted in Figure 5.4(*a*), in order to maximize the economic benefits over one rotation. According to Table 5.2, the undiscounted value of the stand (volume times timber prices minus constant planting costs) increases as long as the trees continue to grow. Still, it is obviously not the right policy to wait until the strand stops growing. When considering the optimal economic time for harvesting, one should realize that allowing a stand to grow for an additional period means investing the current sales value of the stand in the process of growing wood for one more period. In each year the other alternative is to harvest the stand and to invest the money elsewhere. According to our biological model, the tree growth rate, that is, the CAI in Figure 5.4(*b*), starts to decrease when the trees are about 50 years old As long as the CAI of the money invested in growing trees is greater that the CAI of returns on other investments, harvesting must be postponed.

Table 5.2 : Computing optimal rotation.

Age	Vol. (m^3/ha)	MAI (m^3/ha)	CAI (m^3/ha)	Timber Value (ECU)	ROI(%)	Discounted Value (r=3%)
20	4.0	0.2	6.9	600.0	172.3	–4667.8
30	96.5	3.2	11.3	14467.5	11.7	960.4
40	222.7	5.6	13.7	33408.0	6.2	5241.5
50	363.0	7.3	14.1	54442.5	3.8	7418.7
..						
55	432.1	7.9	13.5	64814.I	3.1	7753.3
56	445.5	8.0	13.3	66829.1	3.0	7766.8
57	458.8	8.1	13.2	68817.0	2.9	7763.7
60	497.3	8.3	12.5	74592.0	2.0	7660.7
..						
69	596.7	8.6	9.4	89507.7	1.6	6643.7
70	605.9	8.7	8.9	90877.5	1.5	6477.6
71	614.5	8.7	8.4	92178.8	1.4	6120.1
..						
80	668.8	8.4	3.4	100320.0	0.5	4427.8

The trees should be harvested immediately when the CAI in the value of the stand is equal to the CAI of the returns on the best alternative investment. Why? Because during the next period one already obtains more interest from an alternative investment (for example, a savings account). Table 5.2 column 6 shows how many per cent the value of the CAI is of the total value of the stand. This is the rate of return of investment (ROI), that is, the increment of the value of the stand when letting it grow for another period. If the market interest rate is 3 per cent, the owner should cut the stand at the age of 56 years, because in the next year the stand already yields a lower rate of return than the best alternative investment. In other words, at the age of 56 years the relative value growth of the stand equals the interest rate.

Thus we have found the time for harvesting the stand that maximizes the percent value of net income over one rotation. The net present value of returns is maximized over one rotation by harvesting the stand when the CAI of the value of the stand is equal to the CAI of the sales income of the stand invested elsewhere. Another way to reach the same solution would be to calculate the present value of the net benefits from selling the stand (that is, the present value of net returns minus planting costs) in successive periods. This is shown is Table 5.2 (column7) using the 3 per cent interest rate. This, of course, produces the same answer: 56 years. As the economic solution takes into account the cost of time (discounting), one can assume the optimal economic rotation to be generally shorter than the optimal rotation when the MAI is maximized. This is also the case in our example, in which maximizing MAI provides the rotation age of 70 years whereas the economic solution over one rotation is only 56 years.

An increase in the interest rate shortens the optimal rotation period. The relative growth rate decreases with age. Thus the alternative investment becomes profitable earlier with a higher interest rate. This can be seen in Table 5.2. If the interest rate is, for instance, 5 per cent it will be optimal to harvest the stand before it is 50 years of age, because ROI is 5 per cent somewhere between 40 and 50. Planting costs are paid at time zero. Thus they do not vary with the age of the trees. Whatever the constant planting cost, the harvesting time that maximizes the present value of the returns also maximizes its present value net of planting costs. Of course, if planting costs are higher that the maximized present value of the stand, then it will not be optimal to plant the timber for commercial purposes. For some regeneration measures, this is not at all an uncommon situation in, for example, the northern parts of the boreal forests.

We have solved an economically optimal rotation problem, but this is not yet the correct Faustmann solution because we have neglected one important factor. We did not take into account the fact that after the first harvest the forest can be replanted. For this reason, the rotation period obtained above is longer than the optimal rotation period of the correctly formulated problem. When the first harvest is postponed al successive harvests are also postponed. This delay will cost money if there is a positive interest rate. Therefore, as realized by Faustmann, and infinite cycle of rotations must be considered when (the) economically optimal harvest time for an even-aged stand is calculated.

The other way to reach the correct solution is to realize that forest sites must yield at least as much in forestry as they would yield in other uses. Including the land rent in the costs of growing timber and maximizing the present net value of the harvest over one rotation therefore also leads to the correct Faustmann rotation. It can be shown that the Faustmann rotation is always shorter that the rotation which maximizes the physical timber production (MAI) of forest land.

As presented above the basic problem of forest economics is, of course, simplified in many ways. For example, products and services of the forest other than timber are excluded from the discussion. Should the forest-owner value the non-timer amenities of his forest property, he would take these into account when decision the optimal time to harvest a stand. One can assume that non-timber amenities of a forest are an increasing function of the volume of the stand. Therefore, taking them into account when deciding the optimal rotation age would obviously delay harvesting if compared with the case when only the sales value fo the timber is considered.

Forest Management Planning at the Stand Level

Important parts of forest economic theory are based on stand-level analyses. Stand management optimization is also the core of practical forest management planning. For empirical applications models that realistically describe the development of the stand over time are required. These models (stand simulators) replace the time-dependant growth functions of qualitative models, such as we used in the numerical example above. Stand simulators are means for obtaining a realistic development of the volume of the stand given the tree species and treatment alternatives. In addition to thinning—ignored above—stands consisting of more than one species can be studied. In the 1980s, models simulating the growth of individual trees instead of whole stand were introduced, and raised an even wider selection of silvicultural questions for analysis than did the stand-level growth models that were used before.

In addition to the stand simulator, a mathematical optimization program (algorithm) is required to calculate the optimal treatment policy. In order to determine the optimal time for harvesting in a more realistic setting, this optimization program replaces the calculation that was carried out manually above. The optimization is no longer as straightforward as above. However, the principle remains unchanged. The program searches for the treatment policy (number, intensity and timing of thinnings, and the timing of final harvest) that meets the given objective.

Figure 5.5 presents an example of the output of this type of optimization program. Optimal treatment policies for a Norway spruce stand are presented for four different cases. The thin continuous line represents optimal years, intensity of thinnings (sharp drops in the line) and the year of final harvest if the objective is to maximize (MAI). Three other lines represent the optimal treatment policy if the objective is maximum soil expectation value, that is, maximum present value of net benefits from an infinte cycle of rotations (Faustmann solution) at 1 per cent, 3 per cent, and 5 per cent discount rates. The results show that maximizing MAI gives the longest rotation period, while maximum soil expectation value at a 5 per cent interest rate gives the shortest rotation. In the latter case it is optimal to thin the stand only twice, whereas the other three alternatives recommend three thinnings.

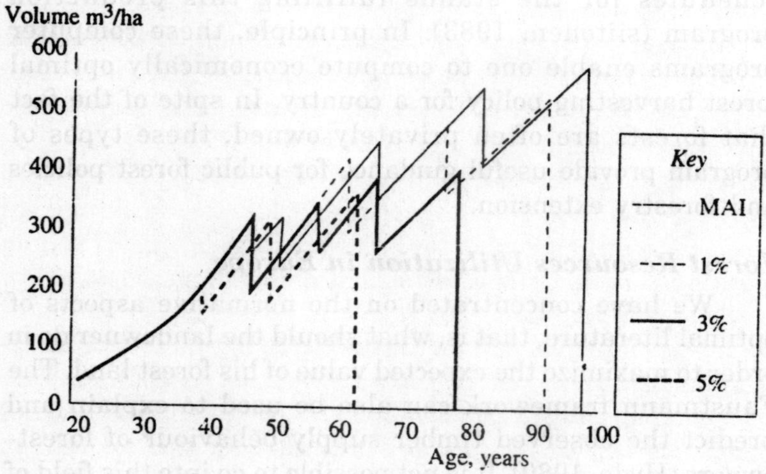

Fig. 5.5 : Optimum thinning programs for different objectives. Optimum thinning and rotation for (1) mean annual increment, and (2) maximum soil expectation value at 1 per cent, 3 per cent and 5 per cent discount rates (Valsta, 1992).

Forest Management Planning at the Forest Level

The other line of research aims at producing optimal treatment programs for forest management planning at the forest level with many different kinds of stands. In a number of countries both private and public forestry is regulated by jurisdiction; forestry extension organizations translate the legislation into accepted management practices. In countries with a long tradition in forestry and forest industries, public forest policies are often based on forest-level optimization models. Under certain conditions the results obtained at the stand level can be applied directly at the forest level.

The primary aim of forest-level models is to provide a system for versatile analysis of the timber production possibilities of a forestry unit (whatever its size). They can be used to find the optimal production program for the forestry unit, in accordance with goals and restrictions provided by the user, and the treatment schedules for the stands fulfilling this production program (siitonen, 1983). In principle, these computer programs enable one to compute economically optimal forest harvesting policy for a country. In spite of the fact that forests are often privately owned, these types of program provide useful guidance for public forest policies and forestry extension.

Forest Resources Utilization in Europe

We have concentrated on the normative aspects of optimal literature, that is, what should the landowner do in order to maximize the expected value of his forest land. The Faustmann framework can also be used to explain and predict the observed timber supply behaviour of forest-owners (Hyde, 1980). It is not possible to go into this field of research here. We note only two things: the above-described optimization methods are primarily meant to be used in, for example, selecting regeneration measures or treatment alternatives for individual stands or in formulating the forest policy of a country. The short-term timber supply, on the other hand, is supply from an existing stock of growing merchantable timber planted, for instance, 50 or 100 years ago. It is affected by various other factors than those accounted for by the optimal rotation models presented above. However, the Faustmann framework also provides a useful yardstick in this case and can be used to derive a *priori* conclusions, for example, concerning the effects of changes in the timber price or interest rate on aggregate timber supply in the short term. However, alternative frameworks for explaining the observed short-term timber supply exist.

Optimal rotation calculations at the stand level assume that the market is willing to accept all harvested timber at given prices. According to Sedjo and Lyon (1992), the conventional wisdom among most foresters is the forest resource is becoming increasingly scarce. The available statistics on saw timber prices (USDA, 1988) indicate a rising long-term trend which seems to support the conventional wisdom. An explanation of this secular trend may be the mining (that is, clear-cutting without planting new trees) of old-growth forests. The above discussed approach to optimal timber production does not imply mining of old-growth forests. Instead, the frame of reference is the kind of managed or plantation forestry having been carried out in central Europe since the 19th century.

Sedjo and Lyon (1992) also mention that all timber prices have not followed a rising pattern. In particular, there seems to be no long-term rising trend in the price of pulpwood. Even the rising trend in saw timber prices seems to have evened out since 1950. The fairly stable pulpwood prices and levelling off of the other price trends may partly be explained by the transition from harvesting old-growth timber to harvesting managed forests. For example, in Finland both timber prices and harvest levels have stayed almost constant over the past 40 years or so. Still the total amount of raw material wood used in Finland after the Second World War is larger than the present standing stock.

The preliminary results of an inquiry conducted by the FAO's Economic Commission for Europe (ECE) actually indicate that the harvest rate in Europe is much below the annual increment of European forests (Pajuoja, 1995). Thus the maximum potential sustainable harvest rate, that is, the rate of harvest that would keep the timber stock constant, is higher than present actual harvest. This, together with the observation that timber prices have not increased or have increased only modestly in Europe during the past 40 years, seems to indicate that the demand for wood is below the present growth of European forests. Of course, sharp increases in the demand for timber from other parts of the world can change the situation dramatically. In Figure 5.6 the expected development of the growing stock of European forests, annual increment and total fellings (harvest) is projected to the year 2040. Figures are based on the answers of country representatives in the above-mentioned study of the FAO/ECE.

The present large growing stock with net annual increment (NAI) larger than fellings (Figure 5.6(b)) may reflect the fact that European forestry has traditionally had goals other than purely economic ones (for example, the *'in situ'* value of growing stock). Alternatively, future demand may not have come up to expectations, or it may be that the interest rates used in forestry are considerably lower than expected in other parts of the economy. Recently other explanations for the substantial increases in European forest resources have also been proposed. For example, Kauppi *et al.* (1992) suggest that the explanation may partly be the fertilization effects of pollution (acid rain), which overrule the pollutants' adverse effects at least for the time being.

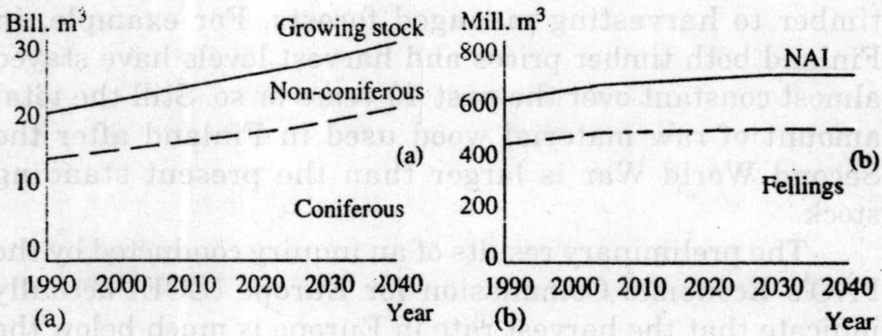

Fig. 5.6 : Future development of European forests. Forecasted future development of the standing stock, net annual increment and fellings (=harvest) in Europe to the 2040.

This situation means that at present there should be no need for an immediate conflict between timber and non-timber services of the European forests. If the forests are managed carefully, so that forest site allocation for different purposes can be made and the environmental problems in forests can be kept under control, European forests should be able to satisfy the needs of both timber production and production of other service of the European economies.

6

Welfare Factors and the Environmental Resource Base

Introduction

Gross national product (GNP) has nowadays become commonly used as an index of welfare in a country. If GNP increases, this is assumed to be a sign that the average welfare in the country increases, and if GNP in one country is greater than GNP in another, then that country is thought to enjoy higher welfare. This notion that GNP is a measure of 'material' well-being is very common and has led to consequences for political decision-making. For example, high growth of GNP is used by the incumbent government as an argument for its economic policies, while low growth is used by the opposition parties as an argument against the present policies. Each government will therefore try to promote the growth of GNP.

One of the conclusions of this chapter is that GNP is a bad measure of wellbeing. However, the focus on GNP has shown without doubt that it is desirable to use a single and simple measure of well-being, in spite of all its shortcomings. It is therefore legitimate to ask whether it is possible to improve the existing measure (GNP) in order to get a more reliable measure of well-being.

The purpose of this chapter is to discuss the theoretical basis for using a national product concept as an index of well-being and the assumptions that are needed for making such an interpretation. However, it should be emphasized that national income accounts have other, probably more important, purposes than measuring well-being.

Economic theory guarantees that if a few technical assumptions (such as convexity of production possibilities and preferences) are valid, then there is a linear index that can be used for evaluating small projects, that is, that can be used for establishing whether a small project will enhance human well-being or not. The notion of a project should be interpreted in the broadest sense. A project is simply any change in the organization of the economy that affects present and future consumption streams. It covers the investment in a pulp and paper plant as well as the aggregate of all changes in the economy during a year! A project is small if it does not affect prices, and projects must be small in order to make construction of linear indices possible. If prices do not remain constant, any index must necessarily be non-linear, because one would then have to include changes in consumers' and producers' surpluses. The established notion or GNP is a notion of a linear index, that is, the sum of (constant) prices times quantities. In trying to extend the GNP measure to something more appropriate and at the same time keep its linearity, it is thus necessary to restrict the analysis to small projects.

The index we are looking for is obviously a marginal social cost-benefit rule, that is, a rule that tells us whether a project (or all projects during a year) is socially desirable or not, that is, ranking the social desirability of projects. However, it is obvious that such a social cost-benefit rule is dependent on the consequences for income distribution. Established national product concepts do not include distributional impacts, and this is why we will disregard them in this chapter. There is no theoretical reason, however, why distributional impacts should be neglected in constructing an index.

The social desirability of different projects can be represented by an index in many different but consistent ways. A unique index, therefore, should not be expected to be the result of our investigation. Basically, the different but consistent indices will differ from one another in the choice of base prices. In particular, the future of the economy is very important in determining the appropriate prices. The probably most natural choice is based on the use of future optimal growth as a benchmark. One index (which we will study in some detail) can therefore be based on prices along an optimal growth path (which does not imply that the economy is growing optimally). Another index can be based on local prices, that is, the prices households are facing when they make their decisions on consumption. One could also construct an index based on local prices for the current period and on optimal prices for the future (which resembles the standard interpretation of current computations of national product). In fact this is perhaps the most natural approach, which is analysed in a forthcoming paper.

In order to illustrate these ideas, let us limit ourselves to an economy with two consumer goods and with one individual. The economy is depicted in Figure 6.1.

X and Y are the two commodities and the curve TT' is the production possibility frontier (that is, the curve that delimits the set of feasible productions: a point below the TT' curve can be produced while a point outside the curve cannot be produced). II' is one of the indifference curves for the individual (an indifference curve is a curve that combines bundles of consumption that the individual is indifferent to) and is tangent to the production possibility frontier at A. This means that point A is the best point for the consumer. The common tangent, pp', defines the optimal prices and we could define an index such as the national product, NP, as:

$$NP = p_x X + P_y Y.$$

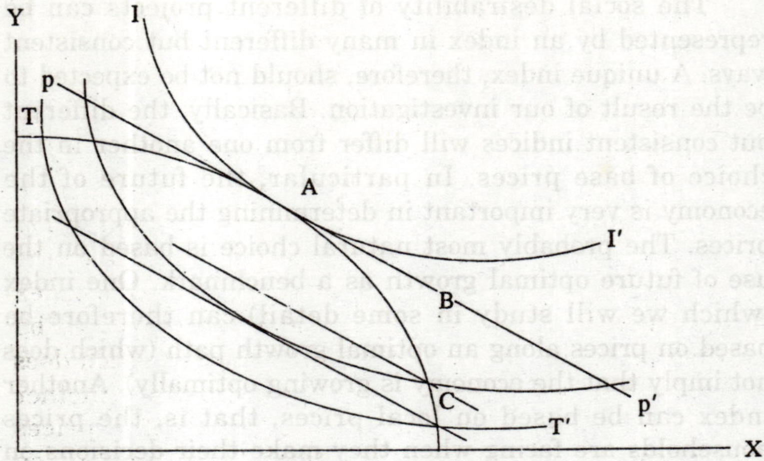

Fig. 6.1 : NP with optimal prices.

Assume that the economy is not in optimum A, was in the non-optimum point C earlier, but has now changed from C to B. Was that an improvement in welfare? We can see from the indifference curves through points B and C that it actually was an improvement. But note that our index, represented by the straight lines parallel to pp' but through points B and C, also indicates improvement. In this situation, the index is very simple. Use the optimal prices, irrespective of whether the economy is in optimum or not, to calculate the national product. Another index, based on local prices is illustrated in Figure 6.2.

Assume that the economy is far from optimum A, that is at point C. Irrespective of the reasons for being in this inefficient situation, the local marginal valuations of the individual are given by the tangent to the indifference curve through point C, that is, of the slope of line pp'. This slope defines the local prices, which may be quite different from the actual prices at C and quite different from the optimal prices. If the economy is undergoing a change from C to B, the national product index, defined with local prices pp', will increase and so will welfare, as can be seen from the indifference curves. Thus the national product calculated with the current marginal consumer valuations is an appropriate index for a marginal social cost-benefit rule.

This latter index also has the advantage that (if the production possibilities and the preferences are convex) if the economy is undergoing a series of changes that all improve the national product, the allocation will ultimately lead towards the optimal allocation. This results from the gradient process, which is a mathematical procedure for searching a maximum. The idea can be compared to climbing a hill by choosing the steepest path. By following such a path, one can be assured that one is following the shortest way to the top.

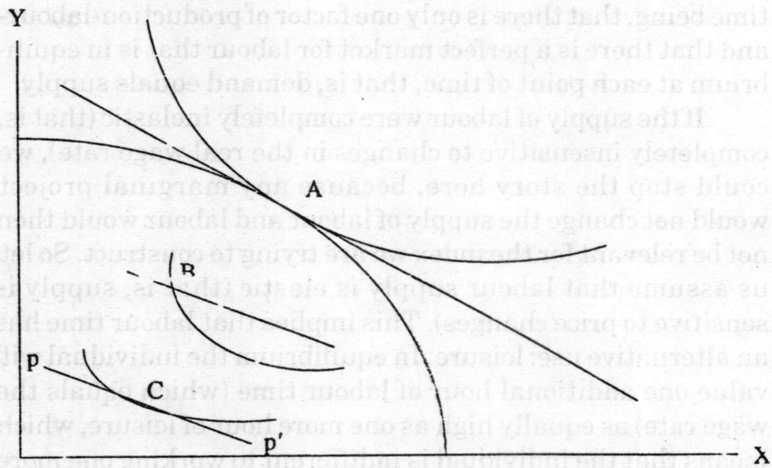

Fig. 6.2 : NP with local prices.

The rather simple ideas explained above will now be used to develop a framework for the construction of an index for a much more complicated economy than we have discussed so far. However, some of the restrictions we have used are not important. The restriction to two commodities is, in fact, more general than is necessary, so we will limit ourselves to one consumer good without restricting the generality of the analysis. We will also inantain the assumption of only one individual, so as to bypass distributional issues.

The issues we consider are the following:

- Labour as a factor of production.
- Non-market flows of services (environmental damage).
- Capital stocks.
- Valuation issues.

Labour

We have assumed the existence of a production possibility frontier. The existence of such a frontier depends on the existence of factors of production and institutions that allocate these factors of different uses. Let us assume, for the time being, that there is only one factor of production-labour-and that there is a perfect market for labour that is in equilibrium at each point of time, that is, demand equals supply.

If the supply of labour were completely inelastic (that is, completely insensitive to changes in the real wage rate), we could stop the story here, because any marginal project would not change the supply of labour and labour would then not be relevant for the index we are trying to construct. So let us assume that labour supply is elastic (that is, supply is sensitive to price changes). This implies that labour time has an alternative use: leisure. In equilibrium the individual will value one additional hour of labour time (which equals the wage rate) as equally high as one more hour of leisure, which means that the individual is indifferent to working one more hour or having one more hour for leisure. But as we are trying to evaluate only marginal projects, income should not be part of our index. A constant-returns-to-scale project that would increase employment, and hence reduce leisure time, and which uses labour only as a factor of production would not increase (or decrease) well-being because the gains from increased production of consumer goods are completely offset by the cost from reductions in leisure. *Thus with perfect, equilibratory labour markets, labour income should not be included in the national product measure.*

This is a rather startling conclusion and we shall modify it. In view of the historical fact that national income measures were constructed to help design economic policies for demand management, one could argue that labour income is perhaps the most central part of the national product. This implies that the national product measure may have at least two different uses and the appropriate concept for one use will in general be quite different from the one for another use! It is therefore necessary to distinguish between different concepts of national product and preferably we should give the different concepts different names. In spite of this, however, we will continue to use national product as the generic name for our index for evaluating marginal projects.

Our conclusion is based on a number of implicit assumptions. The most obvious assumption is, of course, that labour markets equilibrate. Although we are too familiar with unemployment to accept the proposition that labour markets are in equilibrium, we can still construct an index as if they are in equilibrium, as was demonstrated in the introduction. Or we can use local prices, that is, the workers' marginal reservation price (the lowest real wage rate) for taking a job. Assume for simplicity that the reservation price is zero, that is, the worker is willing to take a job as long as he receives something above zero. The accounting price for labour will then be zero and the wage bill, evaluated at this price, will still be zero. This also comes down to the exclusion of labour income from the national product measure.

The latter discussion has some important implications for the discussion on how to treat 'defensive expenditures', that is, expenditures for cleaning up environments that have already been damaged. It has been proposed that conventional national income accounts would register an increase in GNP because of damage done to the environment owing to pollution. The reason for this is that households or governments will hire labour to clean up. This increase in employment will show up as an increase in GNP. Therefore, some economists have argued that defensive expenditures should be deducted from GNP. This train of thought is wrong however. If the economy is in full employment, then the hiring of people doing the clean-up will be offset by a reduction in production somewhere else in the economy. If the economy is not in full employment, then (assuming we are using the local prices equal to zero) the wage bill should be zero, so there will be no increase in GNP. Thus the defensive expenditures are automatically excluded from GNP provided we value labour correctly!

We have assumed that labour time and leisure are substitutes. However, usually the workers are trained and therefore carry human capital. Part of their salaries is therefore a return on that capital. That part cannot be regarded as a substitute for leisure. Only that part of their salaries that corresponds to 'raw' labour will be valued to the same rate as leisure. The return on human capital should therefore be included in our index. However, this may create serious practical problems. How do we separate out the returns on human capital from the rewards of raw labour? So far there have been no ways of doing this. In the rest of this chapter we neglect valuation of labour and leisure and focus instead on environmental resources.

Flow of Goods and Services

So far we have assumed that welfare is determined only by the goods consumed, and that these goods can be bought and sold on markets. However, human welfare is also dependent on many non-market goods and services. A few of these are already included in the Standard National Accounts (SNA). For example, production in the public sector is in general not marketed but included in GNP. However, there are many other goods and services that are neglected. In fact, human life mainly depends on services provided by fundamental ecological systems, many of which are not provided through markets. The inclusion of these services changes the national accounts in two ways: (1) it changes the current flow of goods and services; and (2) it changes the future flow of goods and services through current changes in the resource base. In this section we limit ourselves to a discussion of the current flow of services, returning to the issue of changes in the stocks later.

The current flow of services from natural capital can affect human well-being in two ways. The services can directly affect human well-being, for example clean air, recreational opportunities and direct food gathering, but can also affect human well-being indirectly, for example when they are used as inputs in production processes. We assume that the value of production is already included in the national accounts. This means that changes in these services are measured by the conventional GNP: a change in the quantity of one service will affect the operating surplus or perhaps the use of other inputs (that is, the profit or the expenditures for substitutes). This operating surplus is accounted for in the value added to the production unit. Thus we do not have to worry about the indirect uses of services from natural capital. However, the direct use of these services is not accounted for in SNA.

There is considerable confusion (to which I myself have contributed) about how to include the direct services. This confusion has to do with our definition of the direct services. One example to illustrate this point. Assume an individual lives close to a highway. The project we are considering (remember we are looking for a marginal social cost-benefit rule) is an increase in the traffic. This increase will increase noise and hence reduce the well-being of the individual. As a consequence, the individual will invest in three double-glazed windows and increased insulation in order to reduce the noise indoors. Should these defensive expenditures be subtracted in order to reach a more appropriate national product index? That depends on the way we are measuring the environmental damage! Let N be the level of noise outdoors. As a result of the increased traffic, this level increases from N' to N''. The original level of noise indoors was M' and thanks to the defensive expenditures this level is maintained. Let $D(N)$ be the damage function (the willingness to pay for quietness). One correct way of adjusting the national product index is to subtract $D'(N)(N'' - N') = D(N'') - D(N')$ from the conventional national product. Here $D'(N)$ is the marginal damage, which is used because we are studying a marginal change. Note that this implies that we are using local prices, that is, the marginal willingness to pay for noise reduction in the actual situation (where noise is N'). In this case, we should make sure that all prices used in computing the index reflect the actual marginal willingness to pay for the goods and services. We could also have made use of the optimal prices. If N^* was the optimal noise level, then the adjustment would have been $D'(N^*).\ (N'' - N')$. In this case, all accounting prices should reflect the optimal accounting prices.

Sometimes a different formulation is used where the damage function is $D(M)$. Because of the defensive expenditures, the indoor noise level is unchanged. Thus the national product should be subtracted by these expenditures. This is correct, if the defensive expenditures are approximately equal to $D'(N').\ (N'' - N')$.

However, there is no reason to assume that defensive expenditures are always a good approximation of the true environmental damage. It may therefore be dangerous to use defensive expenditures uncritically as a measure of environmental damage.

But if one cannot use the defensive expenditures routinely as an index of environmental damage, how should one assess environmental damage? During the past 20 years or so, the techniques for estimating environmental damages in monetary terms have developed considerably. However, these techniques have been developed for use in particular situations where a cost-benefit analysis is needed for making decisions and not for routine production of national product estimates. If these rather sophisticated techniques were to be used, the cost for adjusting the national product measures may be prohibitive. Therefore, simple, robust methods of approximation are needed.

The conventional way of including the public sector in SNA may give some clues. The output from the public sector is in general not marketed and therefore assessing the value of that output directly is not possible. SNA has solved the problem by looking at the cost of production in the public sector (mainly labour costs). This is a correct approach if the public output corresponds to the demand. In fact, if our politicians were able to make correct decisions on the size and composition of the public sector, then the cost of production (including cost of capital) in the public sector will exactly correspond to the value of the output. No one will probably argue that our politicians are that wise, but still this approximation of the value of the public output seems to work. Why not use the same basic idea when approximating the value of environmental damage?

This idea requires that there are politically determined targets for environmental policies (ambient standards, emissions targets and so on). This situation is illustrated in Figure 6.3. Here MC is the marginal cost of abating emissions and MD is the marginal damage from the emissions. The total abatement and damage costs are minimized when marginal abatement cost equals marginal damage cost, that is, at point A.

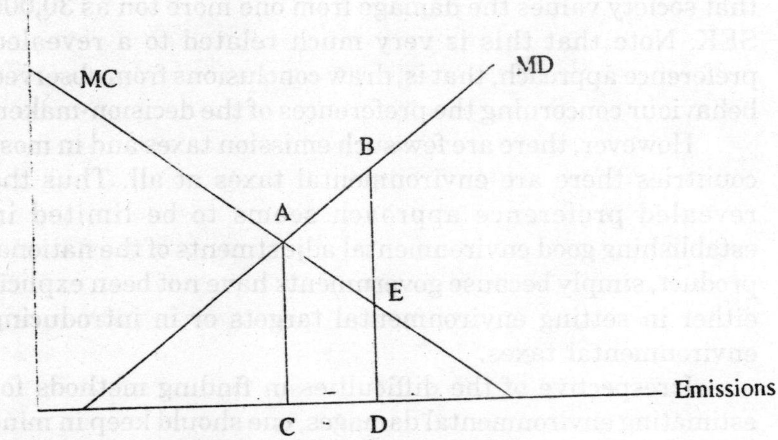

Fig. 6.3 : Shadow prices and government policies.

Assume that the current emissions are at D. Then reducing the emissions to C is desirable. The total cost of doing so is the area ACDE. The reduction in environmental damages from such an emission reduction is ACDB. We immediately see that these two areas are correlated, so that an increase in the environmental damages from an increase in the emissions will be accompanied by an increase in the abatement costs of achieving the optimum level of pollution. Therefore, the cost of achieving the target of emission reduction is, although not a perfect one, an approximation of the environmental damage. If the political system determined-explicit targets for environmental policies, then one could use the costs of achieving these targets as approximations when adjusting (that is, subtracting) the conventional national product concepts for the flow of environmental services. However, politicians seldom determine specific targets in environmental policies and therefore the scope of this approach to evaluating environmental damage is limited.

Similarly, if the political bodies have determined to impose emission taxes instead of setting quantity targets, these taxes can be interpreted as proxies for the social marginal damage from further pollution. In Sweden, the emission tax on sulphur is 30,000 SEK per ton. This implies that society values the damage from one more ton as 30,000 SEK. Note that this is very much related to a revealed preference approach, that is, draw conclusions from observed behaviour concerning the preferences of the decision-maker.

However, there are few such emission taxes and in most countries there are environmental taxes at all. Thus the revealed preference approach seems to be limited in establishing good environmental adjustments of the national product, simply because governments have not been explicit either in setting environmental targets or in introducing environmental taxes.

Irrespective of the difficulties in finding methods for estimating environmental damages, one should keep in mind that almost any arbitrary choice of a reasonable estimator will improve the conventional GNP as a marginal social cost-benefit rule.

The discussion has so far focused on the situation in one country and not included the relations between this country and the rest of the world. These relations include trade and capital movements. However, they also include transboundary pollution. Some of the environmental damage in one country is caused by pollution in other countries and some domestic emissions will create environmental problems in other countries. How should these transboundary effects be included?

Returning to the purpose of environmentally adjusted national product, we need to decide whose well-being should be considered. Should we consider the global population or should we limit ourselves to the citizens of the particular country? Assume that we are interested in a marginal cost-benefit rule for projects in our own country and that this rule should include the impact on the well-being of all individuals, both in the home country and in all other relevant countries. Then we should deduct the environmental damage 'our projects' cause, irrespective of the country in which damage in done. Thus the export of pollution should be counted negatively. Furthermore, as the cost-benefit rule in this case apples to domestic projects only, the import of pollution should not be accounted for. Thus if the objective is to create an index measuring the impact of projects in one country on global well-being, then environmentally damaging activities in other countries should not be included. If all countries based their accounts on this objective, all accounts would be consistent with one another and adding them would make sense. In fact the global national product calculated in this way would give the same result as from a calculation in which we considered the entire globe as one country.

Environmental Assets

The analysis in the previous section dealt only with current flows of goods and services. The discussion is relevant if we are only interested in the well-being in one particular period of time. However, projects during that period will in general have effects on future well-being, through the accumulation and decumulation of productive assets. The opportunities for the well-being of future generations are given by the stock of such assets the current generation is leaving to the future. If we are interested in the well-being of the current as well as of all future generations (and we should be if we take the concept of sustainable development seriously), we should obviously include the net changes in the productive assets in our accounts.

In current SNA, changes in only two such assets are included: real man-made capital and human capital. In GNP gross investments in real capital are included. From the point of view of a marginal cost-benefit rule that relates to the well-being of future generations, this is irrelevant. What is interesting is the net change in the stock of real capital and therefore in the net national product (NNP) where depreciation of the stock of real capital has been deducted. It is obvious that investments that are offset completely or partially by depreciation of the capital stock owing to obsolescence, wear and tear ad so on, will not contribute to higher future well-being.

Changes in human capital are consequences of education, research and development. Some of the costs of these activities are included in the standard GNP measure, but classified as consumption (mainly public consumption). It seems reasonable to reclassify expenditures that increase human capital from consumption to investments. The total investment costs in human capital are composed of the costs of teachers, buildings, teaching materials and perhaps most importantly, opportunity costs for the time students spend on their studies. This latter component is not included in SNA. It seems reasonable that this should be part of the gross capital formation. On the other hand, it seems extremely difficult to define in a theoretically correct way the depreciation charges needed for the calculation of the net change in the stock of human capital.

However, more interesting is the fact that changes in many stocks of natural resources are not included in SNA. As many of these resources generate essential life-supporting services, their exclusion may severely distort the net national product estimates. Trying to include changes in the complete asset base in the accounts should have high priority. How changes in these assets should be included in the national accounts is still controversial, although this should not be so. Once again, starting with the objective of creating a marginal social cost-benefit rule, we have to decide what the impacts of current projects on well-being will be. In order to do this we must make some assumptions on what the future will look like in order to be able to derive the appropriate shadow prices.The simplest and most practical assumption is that the economy will follow an optimal path from now on. Remember the discussion on Figure 6.1 in section 1. There we concluded that we could use the optimal prices even if the economy was not at the optimum. The same holds for this dynamic setting. We use the optimal path as a vehicle to compute accounting prices which then are used to value the current changes in resource stocks. These accounting prices are defined as the value of the increase in future welfare if we saved one more unit of the resource today. It is obvious that it is difficult to compute the prices, but here we will not go into the problems of estimating them. Having computed these accounting prices, it follows that we should add the value of changes of resource stocks at these prices to the conventional national product.

This conclusion has some important implications. First, we include the price times the change in the resource stock. The change in the accounting price is not included, that is, we do not consider capital gains or losses (basically, a capital gain for one individual is a loss for other individuals and they will therefore add to zero). Secondly, for exhaustible resources, the change will be negative unless new discoveries are made. If the discoveries are deterministic (which is perhaps not very likely) in that one can predict with some certainty that new reserves will be found if investments are made in explorations, then the net stock may increase. On the other hand, if discoveries are completely random, then changes in the stock should be valued with accounting prices that take into account the possible (but not certain) future discoveries . These are known as 'contingent claims prices'. However, this is probably only a theoretical speculation as such prices will be extremely difficult to estimate.

For many of the resources included in the stock of natural capital, the same problems will arise as for the flow of services discussed in the previous section. This is exemplified by the stock of biodiversity. There are no market prices for such assets because there is no trade in biodiversity. Instead, one has to follow the same rather *ad hoc* approaches that were discussed in connection with the valuation of environmental services in the previous section. However, there are important resources for which there are market prices. Most of the exhaustible resources are traded on markets and the established market prices could be used as a first approximation in computing the appropriate depreciation charges.

Net National Product

Let us summarize the findings so far. We have developed a marginal, that is, linear index that will show whether the performance of the economy during a specific time period has improved the present value of current and future well-being or not. Such an index may be as follows:

NNP=

consumption of marketed goods
+ public expenditures;
– flow of environmental damages;
+ the value of the net positive change of real capital;
+ the value of the net change in human capital;
– the value of the net decrease in the environmental resource base (+ if the net change is positive).

It is possible to show that this NNP concept has some rather interesting properties:

1. NNP will approximately be the return on the total wealth in the economy, that is, the sum of the values of real capital, human capital and natural capital.
2. NNP will approximately be the maximum constant well-being that, in the absence of non-anticipated technical progress, is feasible. A higher well-being today than NNP will reduce well-being in the future.

The NNP is a linear measure. In fact, it is, if we are following an optimal path, a linear approximation to the optimal present value of future welfare. It can be shown that this optimal present value of future welfare will exactly correspond to the return on the total wealth. However, we may not be following an optimal path, which is the reason for the two propositions above having to contain the word 'approximately'. NNP can thus be viewed not only as an index for marginal cost-benefit analysis, but also as an index for whether the development is sustainable or not. To understand properly this interpretation of the NNP measure, some mathematics is required which is beyond the scope of this volume. However, the basic idea is simple. By sustainable development we mean a time path for the economy, such that welfare is not decreasing. Property 2 above then guarantees that if we limit the well-being today to the level of NNP, we will be able to continue at that level or higher (if there is technical progress) in the future.

An Accounting System

The theoretical arguments put forward in the previous sections are obviously different from the basic idea underlying the SNA. Furthermore, they are also quite different from the ideas underlying integrated environmental and economic accounting (IEEA). Obviously, the ideas expounded in this chapter will be very difficult to implement, particularly because of the inherent difficulties of valuing environmental services. However, it is possible to include the arguments above in an accounting framework similar to SNA. A brief presentation of such a framework is given in this section. The basic building block is a SAM—a social accounting matrix - which can be used to summarize the accounting system in a compact way. Table 6.1 is for illustration only. It does not take into account the conclusion that wages for raw labour should not be included. It does not include international relations, neither does it include a finer classification of institutions, including the public sector, instead of having households as the only institution.

The economy has been divided into seven different accounts: households, two factors of production (labour and capital), one for production, one for abatement (or for expenditures for environmental improvements in general), savings/investment account and one for the environment. Each row corresponds to incomings for the corresponding account, while the columns correspond to the outgoings. The households thus have incomes that corresponds to wages, profits and an implicit rent on the environmental assets. These incomes are spent on purchases of consumer goods, savings (including savings in the form of net additions of the environmental assets) and the imputed value of the environmental services consumed by the households. In a corresponding way, the ingoings to the production account are the incomes from sales to households, sales to firms for investment goods and from the sales for abatement. The total gross savings in the economy consist of household savings and depreciation of capital stocks. These savings correspond to the investment in real capital and the net additions to the stock of natural capital.

The NNP is given as the sum in the first column, that is, the value of consumption plus net savings minus the reduction in environmental services, that is environmental damages. The net national income is given in the usual way as the sum in the first row, that is, wages plus profits plus the return on environmental capital.

Table 6.1 : Social accounting matrix.

	Households	*Labour*	*Capital*	*Production*	*Abatement*	*Savings / Investment*	*Environment*
Households		Wages	Profits				Rent
Labour				Wages	Wages		
Capital				Profits	Profits		
Production	Consumption				Inputs	Gross investment	
Abatement				Abatement			
Savings/ Investment	Household savings			Depreciation	Depreciation		
Environment	Environment	services		Environment damages	Environment improvement	Value of the net change of resources	

Conclusions

The present GNP is not a good indicator of the performance of an economy. It is not even meant for measuring economic performance. However, in view of the demand for such an indication it seems reasonable to try to develop an index that catches changes in well-being. In this chapter we have discussed the conceptual background for the construction of such an index, and found that ideally we would try to estimate the NNP, which would be an index for current welfare (consumption minus current environmental damage) and an index for future welfare (that is, value of the changes in stocks of all assets). However, we have not dealt with the problems of computing NNP and we do not know whether this is feasible or not. There have been a number of attempts to estimate NNP, but they have been rather partial in that they have included a couple of environmental and resource issues. The research in this area is increasing, and we will probably very soon have a much firmer base for drawing conclusions on the feasibility of constructing good indicators of economic performance.

How to Value the Environment?

Introduction

One of the central themes in the field of environmental economics has been the valuation of environmental resources. This valuation usually takes place within a framework of a benefit and cost analysis of projects affecting environmental resources, or in assessing natural resource damages. These resources or 'assets' include air, surface-and groundwater, woodlands, unique natural landscapes, wetlands and so on. These environmental assets provide three functions: ecological system support, 'negative' inputs (sometimes referred to as waste 'sinks') in production processes, for example, groundwater contaminated by industrial waste and amenities which are 'consumed' directly by individuals (clean air, water for household use and recreational services in natural areas). Along with other economic goods and services, all three environmental functions contribute, directly or indirectly, positively or negatively, to peoples' well-being. The environmental functions and therefore the assets which provide them, are in fact also economic goods or services because in modern society they are not free; their provision, maintenance or conservation entails giving up the production of some alternative goods or services. They are distinguished, however, from conventional economic goods and services in that their use does not always involve market transactions. Consequently, explicit market-determined valuations—that is, prices—usually do not exist for them. This is why they are often referred to as 'non-market' goods.

The discussion above implies that if people prefer, say, clean air to polluted air, their welfare would increase if more clean air were provided. However, there does not exist a social institution where these economic desires could be materialized. In the case of other assets, goods or services, economic desires are expressed through a market where the forces of demand and supply interact. Unfortunately, market transactions for an environmental 'commodity' are not possible. The reason for this can be found in the nature of many environmental amenities. They are called 'public goods' in economic theory. A public good is an economic good (or service) which, once provided to an individual, cannot practically be withheld from any other individual who wishes to consume it. Thus once cleaner air is made available for a neighbourhood, anybody living in that neighbourhood can enjoy it. Ever, if one individual were to pay for it, no one else could be prevented from enjoying it. Economists call these non-paying consumers of clean air or any such commodity 'free riders'. There is therefore no market incentive (that is, buyers paying a price to cover the supply costs) for any firm or person to provide these commodities, as in the case of conventional goods, since the supplier would never be able to recoup the cost of providing it; everybody would have an incentive to 'free ride'. Hence markets are not established and the environmental good becomes a *non-market* good. It is nevertheless an economic good because people are willing to pay for it.

The outline of this chapter is as follows: section 2 reviews some of the relevant theoretical aspects which underlie economists' attempts to express the value of environmental assets in monetary terms and the problems and issues which often accompany and sometimes hamper, these efforts. Sections 3 and 4 describe, albeit in rather general terms, the major approaches and their associated techniques for carrying out the economic valuation process. While section 3 deals with the use of indirect market information for valuing environmental assets, section 4 looks at ways to elicit this information in a more direct way. Finally, in section 5 we will discuss what non-economists often marvel about: to what extent do the monetary measures thus derived really figure in the policy-making arena?

Some Theoretical Underpinnings

If markets fail the task of supplying the demand for environmental goods, somebody else will be expected to do it. As is true in the case of other public goods, governmental authorities are usually expected to assume the role of provider, since only a government can collect the necessary payments (for example, through the general tax system or through special taxes imposed on polluters), thus passing on the costs to consumers through higher prices. The task of the government is to help implement the social desires of the community for cleaner air, which would not be provided otherwise. In the absence of markets, how should a government decide upon the optimal level of provision, maintenance or conservation of environmental assets or services? Economists recommend employing a decision-support technique called benefit-cost analysis (BCA).

Benefit-Cost Analysis

BCA is a method of consistent comparison between the advantages and disadvantages of a particular project. Benefit denotes anything that increases well-being. Cost is the opposite of benefit, thus denoting anything causing a loss in welfare. In a world where resources (for example, labour, capital, land and natural resources) are scarce, any project involves a diversion of resources away from any number of other possible allocations. For this reason, the purpose of BCA is not only to compare the benefits and costs of individual projects in order to identify those with positive *net* benefits, but also to identify the project (that is, resource allocation) having the highest net benefit among all feasible alternatives. The choice of the 'best' (that is, highest positive net benefit) project among competing alternatives is logically equivalent to maximizing a decision unit's welfare. Thus if, for a project whose objective is to maintain or improve the state of an environmental asset, welfare gains exceed costs, the project is worth undertaking provided there are no alternative projects capable of making an even larger contribution to net welfare (gains minus costs).

This conclusion is grounded in the branch of economic theory called welfare economics. (A basic postulate of welfare economics is that any change that increases the welfare of some individuals without reducing that of others, is socially desirable and should be implemented:)

Can Money be Used to Value Environmental Benefits?

Yes, money can be used to value environmental benefits. People pay for environmental assets, either directly or indirectly. Directly in that they pay large sums of money to travel to unique natural sites and contribute to organizations that promote nature conservation and environmental causes; and indirectly in that they are willing to pay higher prices or rent for a house in a less polluted or quieter neighbourhood or purchase bottled spring water on the supposition that it is safer to drink than tap water. Economists consider the maximum sum individuals are willing to pay for a stipulated increase in the provision of some environmental amenity (given income level and other relevant attributes) to be a reasonable estimate of its economic value or 'price', thus supposing to reflect the improvement in the level of well-being.

Proceeding in 'proper' economic style, the argument can be illustrated with the aid of a simple example using graphical analysis. Consider a simplified representation of a person's preferences between two consumption options.

One consumption option is represented by his total money income, denoted by *M*, with which he can purchase any combination of market goods and services. His other option is an environmental commodity, say, improved air quality, denoted by *E*. Due to its special nature, air quality cannot be 'purchased' and, if available at all, would be 'provided' at no (direct, out of pocket) cost to any individual by a government authority, such as a local air resources board or a similar entity. Consider a rational individual whose preferences are represented schematically in Figure 7.1 by curves W_1, W_2 and W_3. The curves are drawn in such a way that the level of well-being ('utility in economic parlance) along each curve remains constant, as more and more units of *E* replace units of the other goods and services (that is, *M*) in the individual's overall consumption 'bundle' (or vice versa, as *M* replaces *E* along a given curve). Each curve depicts a different level of well-being, with W_2 denoting a higher level than W_1, W_3 a higher level than W_2, and so on. Higher levels of welfare result from the (potential) consumption of larger quantities of both market goods and services (available through *M*) and the environmental amenity (*E*).

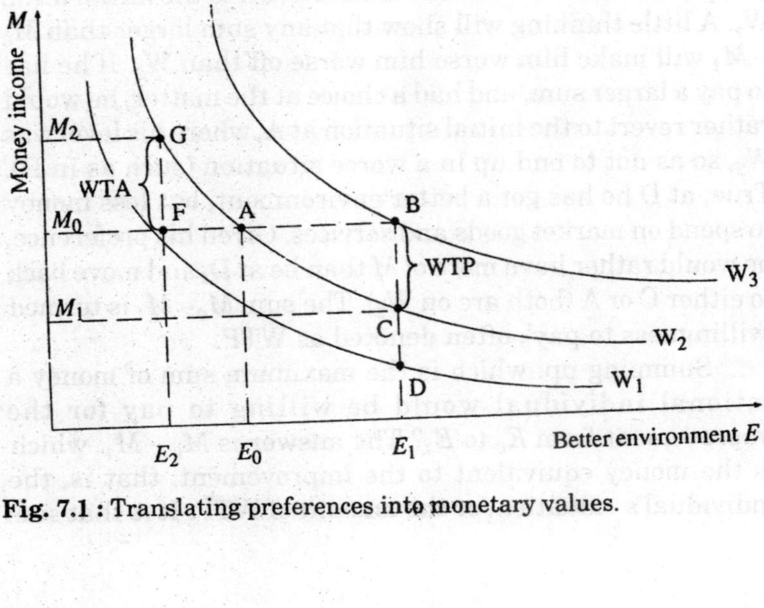

Fig. 7.1 : Translating preferences into monetary values.

Next, assume that the individual is initially at point A, with money income M_0 and environmental quality E_0, which is fixed by the authorities and cannot be directly chosen by him. His maximum level of welfare at this point is W_2, indicated by the welfare curve which passes through point A. Would he prefer the situation at point B to his initial situation at A? Most probably, the answer is 'yes'. Why? Because at B he enjoys more environmental amenity and has the same money income to enjoy whatever market goods and services he wishes to buy with this sum. All this is 'summarized' by the level of welfare associated with the higher welfare curve, W_3, which passes thorough B.

What is this hypothetical increase in E worth to the individual in terms of money? How much-money would he be willing to pay for the above-stipulated improvement in E (moving from A to B)? Remember, this individual is assumed to be a rational person. He should state that at most he would be willing to pay sum equal to $M_0 - M_1$ (reading from the vertical axis). Why ? By paying this sum he will be left with only M_1 to spend on market goods and services, but with an enhanced level of the amenity, namely E_1. This implies that he is now at a new point on his preference 'map', point C. At this point his level of welfare is back down to the initial level, W_2. A little thinking will show that any sum larger than $M_0 - M_1$ will make him worse him worse off than W_2. If he has to pay a larger sum, and had a choice at the matter, he would rather revert to the initial situation at A, where his level was W_2, so as not to end up in a worse situation (such as in D). True, at D he has got a better environment, but less money to spend on market goods and services. Given his preference, he would rather have more of M than be at D, and move back to either C or A (both are on W_2). The sum $M_0 - M_1$ is termed 'willingness to pay', often denoted as WTP.

Summing up: which is the maximum sum of money a rational individual would be willing to pay for the improvement from E_0 to E_1? The answer is $M_0 - M_1$, which is the money equivalent to the improvement, that is, the individual's valuation, or the amount of WTP. It is that sum of money which, when subtracted from the initial income level, keeps well-being constant at its original level prior to the increase in the environmental good from E_0 to E_1.

From this it follows that the benefit for, or loss to, the consumer due to a policy change (for example, moving from one level of the environmental amenity to another) is measured by the amount of money income that must be subtracted or added to leave the consumer's level of well-being unchanged. Likewise, one can analyse the reverse situation (albeit much more far-fetched): what is the minimum sum of money the individual is willing to accept as a kind of a 'bribe' in order to consent to a decrease in E, say to E_2 (left of E_0). Again, after some consideration, one can see that the minimum sum is equal to the distance $M_2 - M_0$. This will move him from point F on curve W_1 back to the original welfare level, W_2, at point G. This sum is called the 'willingness to accept' (WTA). The two concepts, WTP and WTA, provide monetary measures of the worth or value of a non-market good to an individual. For various technical and theoretical reasons, most economists have advocated to use WTP in empirical studies. Whatever the case, if these valuations are summed up for all households of the community, a monetary value for a change in the level of an environmental amenity for the whole community can be obtained. Economists sometimes refer to this value as 'social benefits'.

Beyond Myopic Anthropocentric Values: Valuing Non-use Benefits

Natural resources often provide two different categories of benefits: those associated with active use of an environmental resource (or use which took place prior to damage) and those associated with non-use, or 'passive use'. 'Active use', in the sense employed here, includes both consumptive uses, such as the logging of forests or use of water for drinking and irrigation, and non-consumptive use which include recreational activities that (usually) leave the condition of the resource unchanged. Non-use benefits arise when individuals who make no active use of a natural resource might nevertheless derive satisfaction from its mere existence, even if they never intend to make active use of it.

Take, for example, water quality improvements of a lake, the unique scenic and ecological qualities of which attract visitors nationwide. Benefits from using may involve improved recreational opportunities, better water quality for drinking and irrigation or aesthetic benefits from viewing the (cleaner) body of water while recreating or passing by. Non-use benefits arise when the possible loss of the lake, say due to eutrophication from industrial or agricultural effluent, would be of concern to the general public, including people who have never visited the lake or may never do so. In this case, users and non-users alike may be concerned about the impact on themselves and on others (as well as on future generations) and about the loss of a unique environmental resource. The latter arises from a pure concern for the preservation of an environmental asset for no reason other than maintaining the existence of that asset. The term 'intrinsic value', which is a belief that apart from any source of human utility, natural environments (or species and therefore their natural habitats) have an intrinsic right to exist, has been frequently invoked in this connection.

Since the concept was first proposed by Krutilla in 1967, the proper role of non-use value in economic analysis has been the subject of debate. Many economists (though by no means all) agree nowadays that non-use value is an important component of total economic value defined as the sum of non-use value and use value. They argue that the inclusion of non-use values in damage assessment is essential if compensations are to constitute the proper price signals for resource use. The socially optimal compensation must reflect the loss sustained due to environmental impairment in order to provide incentive for the damager to take appropriate preventative action.

A Few Right and Not-So-Right Caveats

It should be reiterated that the point of this chapter is to show that monetary valuation of a non-market good is, in principle, possible. Nevertheless, some readers would no doubt wish to raise a number of caveats with regard to the WTP valuation procedure. One caveat concerns the implicit assumption that in making valuations people possess all the relevant information regarding the effect of environmental amenities, either negative (for example, pollution) or positive (for example, outdoor recreation) on their welfare. People often have misconceptions about true effects and act upon perceptions rather than on the basis of firm knowledge. For example, in dealing with air quality, it is doubtful that individuals are completely aware of the full consequences of poor air quality on their health and hence on their well-being now and in the future. This qualification, however, is not characteristic of environmental assets only. Is a potential car buyer always, or even most of the time, completely familiar with the various technical qualities of different cars? Or of a stereo set? How often is the decision merely based on hearsay, external appearance or friends' recommendations, that is, on very imperfect knowledge?

An even more disturbing issue to some is a philosophical one, concerning the existence of intrinsic qualities associated with environmental assets which go beyond any attempt to capture them in a single economic index. There are indeed services of environmental assets which probably defy quantitative measurement; but to what extent, if any, would they differ from, form example, the spiritual manifestations associated with the purchase of a religious object which can be purchased at a store? Or the aesthetic satisfaction derived from the purchase of a work of art? All this is not nullified by the fact that the religious or artistic object in question does carry an economic price tag, which captures the value placed upon it by the pious or art-loving buyer.

Nevertheless, even the most fervent advocates of monetization would probably agree that there are a number of weighty measurement issues which are unique to environmental amenities and which cannot easily be resolved, if at all. One obvious measurement issue involves the threat to the very survival of the human race and life on this planet in general. Could any finite value be consistent with the welfare associated with the survival of humankind in the long run, given the possible awesome consequences of environmental episodes with catastrophic consequences? Are there any finite monetary values consistent with long-run sustainability of life on planet Earth? A related point is the issue of valuation being an anthropocentric approach. Is the value placed on an environmental resource by humankind capable of taking into account the moral rights and standing of living captures other than man (that is, the intrinsic value)?

Fortunately, most valuations deal with relatively small, non-catastrophic changes changes in the state of environmental assets, where the advantages of using monetary measures are notable:

1. Money measures provide an explicit and clear expression of the degree of public concern with an environmental issue via the willingness of members of society to pay for the environmental good. In a sense it measures the intensity of the public's preferences and concerns.
2. Monetary measures offer a supportive argument for environmental quality: politicians, officials and the public are used to the meaning of gains and losses expressed in money units.
3. The use of monetary measures make comparisons possible with other monetary benefits and costs arising from alternate uses of public funds.

Using Markets to Value Environmental Assets and Services: The Indirect Approach

One rather widely accepted classification of benefit valuation methods distinguishes between observed, or *indirect* methods, and hypothetical, or *direct*, valuation methods. The former group infers an implicit value for a non-market good from observable, market-determined prices or market goods and services in an indirect way. The latter group, on the other hand, aims at revealing valuations of and demand for, an environmental amenity in a direct way, through consumer surveys where questionnaires are used to elicit valuations from individuals or households.

In general, economists tend to have more faith in market-determined valuations, namely prices which result from the interaction of the decentralized forces of supply and demand. In the absence of government intervention, such market-determined prices are considered 'objective' and 'real' in the sense of reflecting the true valuations of consumers. True, demand and therefore market prices are also subject to manipulation by promoters and advertisements of all sorts and thus do not necessarily reflect consumer preferences. But they are not elicited under artificial conditions, as is the case with some of the direct approaches which have been used to value environmental services.

The observed or indirect approach postulates an explicit relationship between the demand for a market good and the supply of environmental goods. The consumption of the former is postulated to be related in some specific way to the level of the latter. From this demand interdependence assumption, a valuation for the non-market goods is obtained by collecting data on how the demand for the market good changes by shifts in the availability (or the quality) of the environmental resource or amenity. In other words, it is assumed that the environmental commodity affects the demand for some market goods through its impact on the well-being derived by households from consuming the market good. For example, air quality or noise may influence housing prices. Better air quality or less noise is expected to coincide with higher property values in the cleaner or more peaceful neighbourhoos. The impact of pollution on health may affect the demand for medical services (for example, doctor visits, purchase of medication, hospitalization). The extent and direction of observable shifts in demand enables economists through an assortment of econometric techniques, to impute WTP values for environmental goods.

The essentials of the various indirect approaches described below are depicted in Figure 7.2. The curve denoted D_0-D_0 represents a demand schedule for some *market* good (that is, the amount demanded by consumers at different prices). Other things being equal, the lower the price, the greater the quantity demanded of that good. The demand interdependence assumption postulates that given a change in the level of some environmental asset or amenity, the demand for the market good will shift, as shown by the curve $D_1 - D_1$. Now, at each and every price, more will be demanded since better quality induces additional demand for the good. However, as the curves are drawn in a price (vertical axis) times quantity (horizontal axis) space, the shaded area between of two curves measures in money units the benefits gained, that is, the value of the environmental asset or amenity.

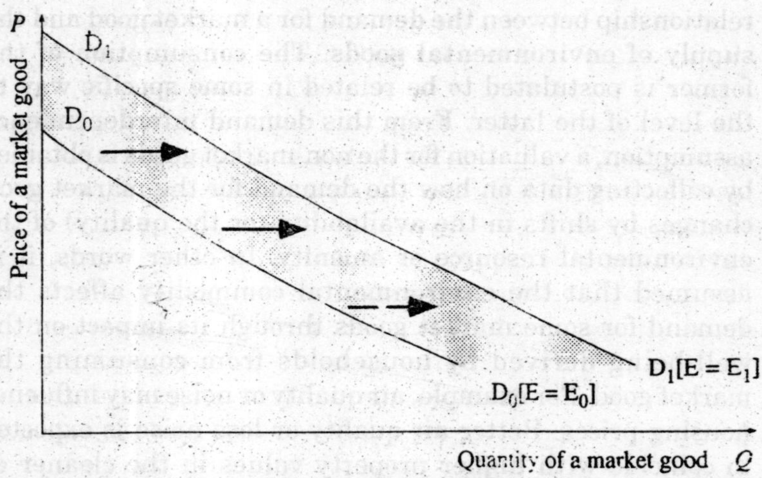

Fig. 7.2 : Measuring environmental benefits from shifts in the demand for a market good.

In this section a non-technical overview of the major indirect methods which have been used by environmental economists to put a 'price' on damages caused by environmental degradation (or, conversely, value the benefits from an increase in the level of an environmental service or resource) is presented. In addition, because of their relevance to the issue at hand, some of the methods used to value reduction in mortality risks (often referred to as the value of human life or the value of longevity) associated with environmental improvements are also presented.

Environmental Amenities and Recreational Services: The Travel Cost Method

The Travel Cost Method (TCM) was the earliest valuation method employed by environmental economists. It has been employed chiefly in studying the demand for and value of, natural resources (national parks, nature reserves and the like) used as inputs in 'producing' outdoor recreation activities. These activities include, among other things, hiking, camping, fishing, boating, swimming and wildlife observation. They are *economic* services produced by the natural resource. Characteristically, they are produced and consumed *in situ:* the recreationist must transport himself to the site in order to experience and enjoy the environmental services provided by the park. This involves expenditure of both money and time. The TCM exploits the fact that consumers reveal their valuation of the natural resource through actual consumption behaviour, in this case travel behaviour. The computation of the value of time spent in recreational activities (for example, wages foregone), the cost of travel (for example, money spent on petrol), and entrance and other site fees constitutes economic value. Basically, it is assumed that the benefits that persons derive from their pursuits within a recreation are directly related to the distance which they are prepared to travel in order to visit it, and hence to their (money and time) cost of travel and other costs. Since this value arises only because people have access to the natural resource when used as a park, it can be considered to be the value of the park, or the implicit 'price' the public would have been willing to pay to secure this form of land use at that natural site.

The TCM was originally employed to attach a value to a natural recreation site whose recreational service, for historical, institutional or cultural reasons, have been provided free or at nominal prices (entrance fees), as has been the case in most countries. The TCM, then, was used to value a limited range of services provided by an environmental resource; that is, those associated strictly with recreation, excluding a range of other environmental services. The impetus for using the TCM to value what was at the time considered 'non-tangible' output, was born out of the need to attach a value to recreational services for a BCA of water resource projects. The more comprehensive BCA involved comparing the benefits from outdoor recreation services with other potential and conflicting, uses of the water resource (for example, damming a river for flood control as against leaving it to flow through its natural course). The desire was to compare explicitly the economic benefits of recreational activities with those of more traditional uses of water resource projects.

During the 30 plus years since its appearance, the TCM has naturally been refined and expanded to deal with more complex situations than in its initial applications. It has often been used to derive values for improvements in the natural resources base, for example, water quality of water bodies within recreational areas. This is done by using the information gathered for estimating the value of a site's recreational services to value changes in environmental assets and the level of amenities provided by them. Figure 7.2 can be used to illustrate the idea. The D_0-D_0 curve would now denote the demand for recreation at a given park, with the vertical axis now representing travel costs, that is, the 'price' of reaching the recreation site. When the quality of a water body in the park is improved, the demand curve shifts to $D_1 - D_1$, as either people who never visited the park would now be willing to incur the costs of visiting the park, or people who visited it would tend to visit it more often. The area between the two curves can therefore be interpreted as representing the money value of the benefits gained.

To summarize: the TCM affords a basis for deriving *use* values of environmental amenities in certain cases, specifically when these amenities are provided by *in situ* natural resources used for outdoor recreation.

Valuing Urban Environmental Amenities: The Hedonic Price Method

In most cases, however, the value of an environmental amenity cannot be revealed via market behaviour connected with traveling expenses to the place of consumption of the service. Other approaches are needed. One of the earliest approaches was the hedonic price model (HPM) which has been used to monetize amenity benefits through the impact of those amenities on prices of related market goods or services. The most common application of the HPM in the environmental literature has dealt with the impact of polluted air on residential housing values. To the extent that property values partly or fully capitalize pollution-induced damages, the HPM may be used to unravel the value of clean air embodied in market prices.

The basic notion underlying hedonic prices is that the price of the related market good (for example, the price of a house) can be decomposed into the prices of the attributes that make up the good. In the case of a house, among its attributes one could count size, age, structural qualities and a bundle of environmental amenities tied to location: the neighbourhood's air quality, noise level, proximity to a waste disposal site or to open spaces, the quality of the school system, and so on. Using econometric techniques, the researcher calculates a hedonic price function, which relates property prices to different dwelling attributes. From this relationship he would derive separate expressions for each attribute, including the environmental ones, showing how price would change when the quantity (or level) of one of the relevant attributes changed. The expression for air quality reveals how much more consumers would be willing to pay for a house in a neighbourhood with better air quality, compared with an identical house in a polluted neighbourhood. This is then the value attached by consumers to improvements in urban air quality in the context of the housing market. Thus, as in the case of the TCM, the HPM exploits the fact that consumers reveal their valuation of the environmental amenity through their actual consumption behaviour in the housing market.

The HPM has been extended to value environmental amenities using wage differentials in the labour market which arise because of variation in urban amenities across cities. Here, the underlying assumption is that, given mobility and time, households migrate between cities in search of, for example, better air quality. Households that value better air quality would be willing to accept lower wages for better air quality. In this case, the researcher would estimate a hedonic wage function in addition to the hedonic property function. The value of the amenity would be the sum of the value component of the attribute in each function. Here, the value of the urban environmental amenity is based on consumer behaviour in two interrelated market goods.

Valuing Human Life: Wage Differentials in the Labour Market

In many of the applications of the HPM, valuation of environmental amenities has been tied particularly (although by no means solely) to the health impacts of air pollution and to how they influence housing prices. There are, however, extreme instances where environmental pollution poses increased death risks. In such cases, policy-makers may wish to focus directly on the benefits from programmes designed to reduce these risks. Here, the pertinent question is what the value is of a human life saved. It is crucial to note that the issue centers on the value of life in a statistical sense of reducing mortality risks, that is, the value of a 'statistical life' and *not* the life of any one recognizable individual. A more appropriate way to phrase the question is, 'what are the benefits of reducing mortality risks from, say, two deaths to one for each 100,000 individuals?' The ethical debate concerning the appropriateness of valuing human life will not be covered here. It should be noted that a range of values of human life based on empirical studies have been employed by the US Environmental Protectin Agency in assessing environmental programmes.

Much of the literature on human-life valuation is based on hedonic market price studies. Basically, these studies have derived statistical life values on the basis of data from labour markets on wage differentials among groups of workers exposed to different levels of fatal or non-fatal occupational risk (see Box 7.1). The theory behind the hedonic price approach stipulates that, as in the case of housing, the price of a job (that is, the wage) can be decomposed into the prices of the relevant attributes that define the job, one of which is the degree of risk involved. Workers in riskier jobs must be compensated with higher wages for bearing additional risks. If, for example, the wages of workers in job A, which is riskier than job B, are higher with an average of \$100, with mortality risks being r_A= 2/1,000 = 0.002 and r_B= 0.001, then the implied value of one statistical life is \$100/ (0.002 – 0.001) = \$100,000.

Although the idea behind this approach appears simple enough, there are a number of serious methodological issues which lessen the universal applicability in the environmental arena. The first problem is the issue of transferability. Can inferences based on valuation derived from the labour market, where risks by and large are voluntarily accepted (given sufficient monetary compensation), be used to assess situations where risks are mostly involuntary, as is the case with exposure to environment pollutants? A second major problem is identical to the one encountered in the housing hedonic model. Wage compensation is a true representation of workers' valuation of mortality risks only if they are fully informed on these risks, a doubtful proposition in many instances.

Box 7.1 Wage-risk studies and statistical life values

Studies attempting to derive estimates of the value of a statistical life from wage-risk differentials have been carried out for the past 15 years; most of them in the United States and a few in Great Britain (at least, as reported in the literature).

Summarizing the more recent studies, after adjusting them to make the result commensurable as much as possible. Fisher *et al.* (1989) report a range of estimates for the mean value of a statistical life of between $1.6 million to $9 million (in 1986 US dollars); that is, they fall within one order of magnitude. In fact, with the exception of a couple of high values, most of the studies yield mean estimates within an even smaller range: $1.6 to $4 million.

Comparable estimates for a statistical case of *morbidity* are few. A figure of $883,000 has been reported for the value of a statistical case of chronic bronchitis, and $200,000 per case of chronic disease.

Valuing Health Damages form Incomplete Market Behaviour: the Cost of Illness Method

The cost of illness method (COIM) has often been employed to estimate economic gains from improved health. COIM estimates the changes in private and public expenditures on health-care and the value of lost production on the basis of the relationship between excess morbidity or mortality and ambient pollution levels (usually referred to as 'dose-response functions')(see Box 7.2). These expenditures, however, do not necessarily measure everything that households spend, or are willing to pay, to avoid poor health. Typically, reductions in pollution-induced health effects result in savings in four cost categories: (1) out-of-pocket expenses associated with illness or injury; (2) foregone earnings or the imputed value of restricted activity days; (3) avertive or defensive expenditures owing to pollution, for example, the cost of installing and operating air conditioners, time allocated to increased physical exercises or the cost of moving to a cleaner neighbourhood; and (4) the disutility associated with pollution-induced illnesses, for example, pain, stress, anxiety and so on (sometimes referred to as the psychological cost of pollution). By its very nature, the COIM overlooks the last two categories: preventive expenditures and the psychic disutility associated with health damages. This leads to underestimating the full burden of pollution on individuals, because COIM estimates tend to reflect only part of individual experience, that is, the first two components of the aforementioned benefits of pollution abatement.

Box 7.2 Dose-response functions

A dose-response function relates ill-health or death to the level of pollution, as well as other intervening variables, such as smoking, age and so on, which can be regarded as being likely to affect ill health. These functions are typically estimated from relatively large data bases, in which (unlike clinical studies) it is seldom possible to control for these intervening variables. The estimation therefore requires rather sophisticated statistical techniques in order to delineate the impact of pollution. Although estimating dose-response functions is usually carried out by epidemiologists and public health researchers, the pioneering study relating air pollution to mortality was carried out by two economists. They found that a 1 per cent reduction in air pollution (sulphate and total suspended particulate matter, or TSP) and all other variables held constant, would give rise to a 0.12 per cent reduction in mortality. They then proceeded to use these relationships in order to attach a monetary value to health damages, using the COIM approach.

In a related study, Ostro (1983) used a large data base from the United States to carry out a regression analysis relating morbidity (population illness ratio) and variables such as race, marriage, income, social status, sex, cigarette consumption, work status, temperature, rainfall and population density. His analyses show that, *ceteris paribus*, a 1 per cent reduction in TSP could lead to a 0.45 per cent reduction in 'working days lost' (a common measure of overall morbidity) and to a 0.39 per cent reduction in 'restricted activity days' (another common measure of overall morbidity, which captures also the impact on the non-working segment of the population).

The reason for this serious shortcoming is that the COIM rests on the premise that improvement in well-being is synonymous with maximizing society's output of goods and services, as commonly measured in national income accounts by 'net national product' (NNP, which equals 'gross national product' - GNP - minus depreciation) or its equivalent — national income. But this is not necessarily synonymous with society's welfare or well-being, which are what it aspires to maximize. Goods and services certainly contribute to our well-being, but so do environmental assets (clean air, for example) which do not enter these output measures. In the same vein, the COIM overlooks sick-day losses owing to pollution of retirees, housewives and children, because they do not belong to the work force and as such their bed-days do not reduce society's measured output (that is, GNP). However, their welfare is nevertheless affected by pollution. Thus COIM is not considered to be a satisfactory method for estimating WTP. On the positive side, however, COIM is intuitively appealing and data on earnings and medical expenditures are relatively easy to obtain. This has led to its popularity among researchers, economists and non-economists alike.

Another problem with COIM is that households do not usually bear the full economic consequences of work-loss days due to morbidity (unless it permanently incapacitates future productivity and leads to a permanent reduction in earnings). This is due to the fact that sick-leave coverage is rather commonplace and medical expenditures are covered by health-care insurance, which is often heavily subsidized. Moreover, people do not normally possess the information which would enable them to assess fully the economic impact of exposure and disease and hence incorporate this information in determining the level of medical expenditure. Pollution-induced outlays are largely borne by society and would be poorly reflected in the WTP of the individual household (as implied in Figure 7.1). Thus *on this count,* the COIM produces higher values than the rest of the indirect methods which, as we have seen, tend to deduce value estimates from market behaviour of households.

Valuing Environmental Amenities as Inputs in Producing Health

The health-production-function method (HPFM) studies household behaviour with regard to the household's desire to improve its well-being, specifically through the health status of its members. Improved health contributes to well-being on four counts: (1) it reduces expenditures on health-care services, thus increasing the amount of time of income left for utility-enhancing goods and services; (2) it reduces the amount of time needed to be spent on preventive or avertive activities; (3) it further diminishes the negative impact on welfare when income is not reduced by loss of work days or not increased due to health-related productivity gains; and (4) it reduces the disutility associated with being ill. To this end, the household consumes an array of market goods and services, including health-care goods and services. For example, households attempt to rectify or prevent health damages associated with air pollution by installing air-conditioners or air filters, by refraining from outdoor activities during high-pollution periods or by buying medicines for the treatment of pollution-induced respiratory ailments.

Better air quality, however, reduces the need for, or use of, these commodities, that is, it may be considered a 'substitute' commodity. The HPFM treats the environmental amenity, market goods and services (including health-care goods and services) as 'inputs' in a process of 'producing' health (hence the use of the term 'health-production-function method'). Improved environmental quality thereby decreases the reliance on preventive and curative 'inputs', that is, it reduces the demand for the other *market* inputs. In applying the method, the investigator uses data on household behaviour to estimate a health production function. From this function the researcher derives the implied trade-off between an increase in the level of the relevant environmental amenity and expenditures on other health-related inputs. For example, other things being equal, better water quality may reduce the occurrence of digestive tract illnesses and thereby the consumption of health-care services which entail both private and social money outlays. Thus HPFM allows the investigator to calculate the implicit value of changes in the provision of non-market environmental assets through the indirect effect on the consumption of market-priced health inputs.

Although seemingly similar, HPFM and COIM are methodologically distinct. It is not possible to go into detail here; an example should suffice. While taken into account by the HPFM, COIM overlooks *preventive expenditure,* money spent by a household to ward off the effects of a deteriorated environment. For example, an asthmatic person may install an air-conditioner as a means of protecting himself against the bad effects of air pollution.

Values of a statistical life have also been obtained from a method which belongs to the HPFM genre. It has often been referred to in the literature as the 'averting behaviour approach'. The method focuses on market behaviour with respect to the purchase of goods and services which may reduce the risk of death, for example, seat belts or smoke detectors. Here, other things being equal (for example, the consumption of all other goods and services, income, and so on), the value of a small reduction in mortality risk provided by the device is measured by the value of the device. The preventive devices, or risk reducing services, can also be viewed as inputs in the process of producing more 'longevity' (in the statistical sense, of course).

Finally, as an aside, a little reflection will show that the TCM can also be viewed as a type of HPFM, a *recreation* production-function model, where travelling expenses, equipment and environmental amenities are combined by the household in 'producing' recreational services at natural sites.

Using Hypothetical Markets to Extract Valuations : The Direct Approach

There may be situations when the data necessary to apply any of the indirect methods discussed in the previous section are either lacking or difficult to obtain. More importantly, there is the case of non-use benefits, which are not associated with the actual use of an environment asset or consumption of environmental services and therefore cannot be easily coupled with the consumption of a market good. It is widely thought that these benefits cannot be valued though any indirect method.

In the absence of appropriate data or interdependent market goods, the alternative is to ask people directly what their willingness to pay for a change in environmental quality is. That is, in terms of Figure 7.1, to ask people to reveal what they think is the monetary value of moving from A to B (or from a to F). This is accomplished by the contingent valuation method (CVM). The CVM invokes a framework of a hypothetical or *contingent* market, through which it seeks to elicit valuations directly from individuals.

A crude form of the CVM was first by Davis in 1963 to estimate the benefits of a recreation area in Maine. Since then, CVM studies have queried people about their valuation of a wide spectrum of benefits from environmental improvements, ranging from improvements in water quality for recreation, reduction in air pollution levels (or the severity and frequency of the respiratory ailments associated with them), reductions in mortality risks from hazardous wastes and improved visibility in national parks, to non-use benefits of reductions in acid rain damages to fish and wildlife or of preserving endangered species and biodiversity.

There are two primary assumptions of the direct valuation methodology. The first is that the consumer is the best judge of his best interests and the second is that the consumer's ability to rank preferences is both rational and knowledgeable. Therefore, special attention must be paid to the design of any CVM study because the hypothetical nature of the CVM methodology could result in uncertainty, doubt and even irrationality in the average consumer when confronted with an imaginary commodity. Attention must also be paid to the constructed market. The 'market' is composed of the commodity to be valued, the institutional context of its provision (for example, national or local government) and the means of financing it, or the so-called 'payment vehicle' (general taxes, local taxes, one-time versus annual payments, a charge on utility bills, entry fees and so on). Using questionnaires and surveys, individuals are induced to volunteer information regarding the maximum sum of money they would be willing to pay for the non-market good, as if they were going to 'purchase' it in the hypothetical market.

Although the CVM has made substantial inroads during the past 20 years into mainstream environmental economic research doubts remain with respect to the reliability of direct valuation, mainly because of the presence of strong hypothetical underpinnings. Nonetheless, research by environmental economists, as well as sociologists and psychologists, into the validity and reliability of the CVM has so far indicated quite encouraging results *vis-à-vis* other methods of preference measurement. There have not been many studies so far which have compared hypothetical and actual WTP (see Box 7.3). the overall impression from those which have is, however, that in general they did not report on significant or unexplainable gaps between WTP values yielded in these two cases.

Box 7.3 CVM versus market (indirect) valuations

In theory, at least, the CVM should produce the same valuations as any other method purporting to measure WTP. Given the hypothetical nature of the CVM, however, an increasing number of studies have attempted to assess its reliability and accuracy against observed behaviour, usually from parallel indirect market studies. Comparisons have most often been made between the CVM and TCM, HPM or HPFM (Cropper and Freeman, 1991; Graves, 1991; OECD, 1989; Shechter, 1992). Although the numbers tend to diverge widely in the case of WTA, results for WTP, on the whole, are highly encouraging and bestow a substantial amount of credibility on the CVM, especially in the case of non-use benefits (NOAA, 1993). Naturally, the degree of correspondence hinges on several crucial parameters, such as the elicitation technique used in the CVM and the nature of the commodity being valued.

A couple of attempts were made to compare hypothetical and actual WTP in an experimental setting. These studies, which dealt with private goods, strawberries and hunting permits did not detect any statistically significant difference between the mean values of WTP under the two experimental settings.

Yet the hypothetical nature of the CVM, and the fact that it depends on sample survey data, could weaken the assumption of rational and knowledgeable consumers and give rise to a number of biases. Biases occur in CVM studies for several reasons. For example, the scenario might contain incentives for respondents to misrepresent their true WTP amounts in order to influence the final outcome (strategic bias), or it may inadvertently contain cues which might subconsciously affect WTP responses. in order to enhance the reliability of a CVM study, environmental economists have been working on developing guidelines and protocols for bias reduction and detection in CVM studies. Thus researchers have stressed that the key scenario elements must be understandable, meaningful and plausible. The WTP questions must be clear and unambiguous. Respondents should be familiar with the commodity to be valued. Ideally, of course, respondents should also have had prior valuation and choice experience with respect to consumption levels of the commodity, so that respondents will have reasonably well-formed values for the commodity, although the likelihood that this condition will be met is rather small. All in all, the investigator must do his utmost to prevent potential response biases.

Finally, the CVM has played a major role in the debate on non-use benefits. In using the CVM in this context, people are almost invariably queried in a direct way to state their WTP for non-use benefits, sometimes for complex issues such as ecological balance, biodiversity, the preservation of tropical forests, protection of endangered species and the like. All the methodological problems raised in connection with the use of the CVM, which have been discussed above, arise with extra force in this case, owing to the difficulty of accurate, commodity description, lack of familiarity with the commodity being assessed, and the lack of experience with anything remotely approaching such commodities in actual markets. Nevertheless, very recently the CVM has been favorably reviewed by a panel of highly respected economists, co-chaired by two Nobel laureates.

From Number to Policy: What is the Vlaue of Monetary Values?

The preceding sections have been devoted to explaining how to measure changes in environmental quality in monetary units. Since policies involving environmental improvements and environmental regulation entail costs, much debate has understandably focused on the expected benefits as the rationale for spending on such policies (see Box 7.4). Although all the benefit valuation methods presented use a common yardstick, that is, money, there is no unique correct method; rather, each has its advantages and disadvantages. The refinement of existing techniques and the development of new ones to accommodate a wider range of values is major challenge to economists.

However, despite advances in valuation made over the past 30 years or so, a serious gap remains. The methods have not been routinely incorporated into the decision-making process. This raises the question: if the product is so useful, why isn't fully utilized? After all, it is often argued that what politicians understand and grasp best in numbers, particularly when it comes to dollars and cents. While it is not possible to provide complete answers, one could nevertheless offer a few explanations which may be of some help in finding ways to remedy this situations and close the gap between academic studies and practice. This may help to find ways to convince decision-makers that in spite of the still many 'holes', there is a substantial amount of cheese mass which it is worth taking a bite from . . .

There is a strong resistance to attaching money values to intangible environmental benefits, arising from aesthetics, preservation of wildlife, human health and the like. It is often asserted that: (1) these are priceless and hence carry no price tag, or (2) that on moral grounds no such tag should ever be attached to them. The implication of the first objection, in economic terms is that their value is infinite; that is, no expense should be spared in saving be spared in saving them from impending damage or for rehabilitating them. This is erroneous. In a world of limited resources relative to human wants, there is indeed a limit to what society can spend on these uses, important and valued as they may be. Environmental benefits have a finite price, that is, value, which may be measured.

Box 7.4 Economy-wide benefit assessments

Although the text focuses on the micro aspects (that is, those dealing with the individual economic decision unit, such as a person or a household of the economic theory which underlies the valuation methodologies, it is the *aggreagte*, or economy-wide benefits and damages which are most relevant in environmental decision-making.

Several attempts to derive such national estimates have been made in recent years in a number of countries, especially the US and western Europe. Thus it has been estimated that for the United States throughout the 1970s, air pollution control policies have yielded benefits (That is, avoided damages) in the range of $9 to $92 billion (1984 US dollars), with the most reasonable point estimate being $37 billion (Portney, 1990). These benefits resulted from improvement in both stationary and mobile emissions and relate to reductions in health, material, agricultural and aesthetic damages. For 1978 alone, the benefits from both air and water pollution control amounted to 1.25 per cent of GNP. Furthemore, benefit-cost analysis has received a substantial boost from the 1981 Executive Order 12291 and has been carried out in its rulemaking process by the Environmental Protection Agency (EPA) for all major environmental regulations.

Estimates of national damages from environmental degradation have also been carried out in a number of European countries (reported in OECD, 1989; Navrud, 1992). In the Netherlands, for example, 1986 air, water and noise pollution damages represented 0.5-0.9 per cent of GNP. This is probably an underestimate, since in reaching this figure it was not attempted to monetize all known damages. A parallel study in Germany yielded a rather high estimate of 6 per cent of GNP for 1985.

It should be noted that economy-wide studies (as reported, for example, in Kuik, *et al.*, 1992) are also concerned with the impact of regulatory programme outlays on macroeconomic variables, for example, GNP, price level and inflation, unemployment, and so on.

It is an altogether different argument to say that there some things in life which should not be valued in money terms, for moral, philosophical or other reasons. That is to say, people simply refuse to associate them with 'any monetary yardstick', period! Although this is perfectly defensible stand, one should be fully aware of its implications in the present context. Decisions on allocating scarce public or private funds for these purposes must be made using *some* criterion, whether explicitly or implicitly, for ordering the 'worthiness' of competing uses for these funds. Any decision which allocates funds to the preservation, rehabilitation or improvement of environmental assets and services necessarily takes them away from other presumably no less deserving uses. Many supposedly 'money-value free' prioritizations in the policy decision-making process tacitly attach price tags to the selected choices (for example, the cost of injury and property damage due to accidents which could have been avoided by building an overpass across a busy intersection).

Policy-makers and the public at large have often been preoccupied with the question of who stands to benefit from, and who would bear the costs of a change in environmental quality. Economists have been slow to incorporate distributive concerns into benefit studies. For example, in interpreting the results of a typical valuation study, say, of an air pollution abatement programme which results in positive net benefit to the community as a whole, the numbers only say the following: regardless of who bears the cost of abatement (consumers through higher prices, producers through reduced profits, or workers through lower wages or unemployment), or who stands to reap the benefit (the people who live in the polluted neighbourhoods), a significant reduction in pollution levels would be a rational policy imperative. Unfortunately, many empirical valuation studies have neglected to address distributional issues and though understandable on purely methodological grounds, this tends to reduce their effectiveness in terms of policy-making. It is essential in many cases to address explicitly the thorny problem of the distribution, or incidence, of benefits and costs among different groups or sectors, that is, who stands to benefit from and who would bear the costs of, pollution abatement.

There may be other reasons for the lukewarm response to valuation studies which have nothing to do with the studies *per se* (OECD, 1992). Political considerations and the bureaucratic framework will surely influence the prospects of implementing the policy prescriptions of valuation studies. It seems that from the point of view of successful impact on policy, what counts most is the relevance of a valuation study to an issue which is high on the political agenda at the time of the study's appearance (as well as, of course, the quality of the product!). These circumstances are conducive to another essential element gaining access to the media as an effective means of reaching the public and policy-makers. Media exposure (let us hope, undistorted) seems sometimes to be a crucial factor in this respect.

Finally, as is the case with similar analyses which cause the decision process to become 'transparent', a well-conducted study of environmental-benefit valuation involves explicit presentation of all the assumptions, parameters and the manner in which decisions have been arrived at. This may be an anathema to those who would prefer a degree of fuzziness, feeling more secure with an 'opaque' process. Some simply dislike results which run counter to what they believe *a priori* should be the preferred alternative. In ancient times they might kill the messenger that bought undesired news; today they shelve the study . . .

Acknowledgements

I am deeply indebted to Shirra Freeman and Elise H. Golan for very constructive suggestions and for a painstaking editorial job on earlier drafts. I benefited from helpful comments and suggestions from some of the participants of the authors' workshop in Wageningen, especially Henk Folmer, Tom Zylicz and Jean-Philippe Barde. Hans Opschoor offered a number of very constructive editorial comments on two earlier drafts. I am, of course, solely responsible for the final version.

□□□

Index

A

Abate acidifying emissions 78
Abate smoke
emissions 58, 59
Abatement equiment 40, 58
Abstract decision
mechanism 56
Accounting System 183-185
Acid rain 3
Adaptive control 125-129
Adaptive systems 90
Aggregation mechanisms 56
Air pollution 24, 215
Antractic
Treaty System 99
Anthropocentric
approach 196
Anti-acidification Policy 12
Anti-pollution
equipment 97
Antibiotics 23
Arbitrarity 20
Arbitrary choice 178
Attractor 83
Autocorrelation 115
Automobiles 32
Averting
behaviour approach 208

B

Benchark 32
Benefit-cost analysis 189
Bifurcations 81
Biodiversity 91, 182
Biological model 143
Biological Productive 2
Biosphere 1
Black box 14
Brundtland Commission 4

C

Capital stuffing 151
Carbon doxide 9
Ceteris paribus 54

Chaotic trajectories 86
Chernobyl 88
Chlorofluorocarbons 14
Choice of
  Instrument 78-79
Clean technology 18
Coase, Ronald 33
Coase Theorem 33-37
Cobb-Douglas
  Production 105
Co-evolving 90
Commodity 139
Competitive market 41
Congressional
  Budget Office 99
Consumption equivalence
  approach 97-98
Contingent
  claims prices 182
Contingent market 209
Convex set 48
Cooperative
  equilibrium 61
Coordinated aproach 5
Coordinating
  arrangement 20
Correcting for
  externalities 25-27
Copper mine 108
Cost-effective
  abatement policy 30
Cost of centralization 77
Cost of cleaning 44
Cost of illness
  method 204
Cost structure 27
Counter intuitive
  propositions 32
Current annual
  increment 157

D

Decentralization
  Principle 76-78
Decision-making 55
Degradation 2
Depletable
  externalities 33
Deplorable resources 133
Discounting 94-112
  Approriate rate 98-100
  consumption
    equivalence 97-98
  control theory 108-109
  Difficulties in
    the method 107-108
  Ethical issues 94-96
  Guidelines 104-107
  optimal growth
    theory 100-102
  sourcing 96-97
Divergent trajectory 107
Dose-response
  functions 204
Dynamic optimization 103

Dynamics 82-93, 148
catastrophe theory 88-90
chaos theory 86-88
Evolutionary system 90-92
Introduction 82-86
Thresholds 92-93

E

Ecological balance 212
Economic-ecological system 90, 109
Economic interaction 19
Economics of Environmental policies 10-12
Economics of fisheries 142-143
Economics of forestry 153-166
Approaches to optimal 155-157
Economically optimal 158-161
Forest management 161-164
Forest resources 164-166
Eco-system management 117-121
Emission permits 43
Emission sources 30
Environment and Natural Resources 7-10
Environmental amenity 192, 199-201
Environmental assets 179-182
Environmental commodity 191
Environmental decison-making 76-79
Envionmental degradation 2, 5, 95
Environmental-economics system 90
Environmental externalities 5
Environmental infrastructure 18
Environmental policy 7
Environmental resources 2
Environmental Technology 15
Environmental utilization space 1, 16
Equilibrium price 44
Ethical issues 94-96
European union 6
Eutrophication 194
Evolutionary systems 91
Exhaustible resources 2, 108
Externalities 8, 17, 22-24, 55
Externality problems 79
Extract valuation 209-212

F

Finnish Lapland 135
Fish population 145
Fish species 84
Fishermen 147, 148
Fishery 150-153
Forestry 9
Free disposal 110
Free riding 60
Full utilization 144

G

Genetic algorithms 92
Global ecosystem 110
Global warming 3, 8
Government Decision
  Rules 62-76
  Decision making 68-74
  Ideal public decision 65-68
  Idealized public
    decision 63-64
  Political economy 74-76
  Repregentative
    democracy 71-72
  views on
    the government 62-63
Government
  behaviour 56
Government
  intervention 62
Grand canyon 24
Green house gases 9
Gross national
  product 167

H

Harvesting policy 144
Hazardous waste 45
Health-care
  insurance 206
Hedonic pricing
  model 52, 201-202
Hedonic property
  function 202
Homogeneity 55, 60
Hotelling rule 124
Housing market 202
Human well-being 175
Hypothetical
  markets 209-212

I

Imaginary commodity 210
Incomplete market
  behaviour 204-208
Individual
  transferable quotas 152
Industrial landscape 15
Integration 84
Internal market 6
Internalized 77

Internalizing
the externality 27-29
Invisible hand 22
Irreversible dynamics 92

K

Keynes-Ramsey rule 104
Krill fishery 127

L

Labour 172-174
Lagrangian
multipliers 103
Laplanders 34
Laundry industry 49, 59
Leviathanian view 62
Liberty 94
Limits to growth 137

M

Macro-economic policy 18
Marginal abatement 177
Marginal
consumer valuation 171
Marginal cost 27
Marginal externalities 30
Marginal product
of capital 101
Marginal
reservation price 173
Marine resources 117-121
Market-determined
Prices 188-197
Market economy 19, 47, 54
Market equilibrium 22
Market integration 80
Market transactions 25
Marketable
emission 37-41
Markets and
Externalities 7
Mill, John Stuart 134
Mineral reserves 1
Misallocate resources 32
Monetary measures 196
Monetary
valuation 113
Monetize amenity 201
Money-value free 214
Monopolist 45
Morbidity 206
Myopic anthropocentric
values 193-194

N

National Product Index 170
Natural
resources 1, 3, 134-136
Net annual increment 166
Net national
product 180, 182-183

Non-convexity 42, 47-51
Non-co-operative
 equilibrium 58-59
Non-linear optimal 126
Non-linear
 system 82, 115
Non-market goods 188
Non-renewable
 resources 1, 124, 136-137

O

Office of Management
 and Budget 98
Opaque process 216
Open access
 fisheries 146-150
Optimal accounting
 prices 176
Optimal harvest 108
Overcapitalization 150
Over harvesting 151

P

Pareto, Wilfred 53, 58
Perfect market
 economy 54
Perverse incentives 10
Physical timber
 production 161
Pigouvian
 taxes 29-33, 37, 44
Plantation for
 forestry 165
Policy-making 6, 13, 215
Polluters 10, 60-62
Pollution 62, 133
Pollution control 51
Pollution havens 6
Private
 externalities 24-25
Private sphere 55
Probabilistic
 techniques 114
Process of decoupling 17
Production-activities 106
Productivity 1, 100
Profit-maximization 140
Protectionism 111
Public goods 51, 188

R

Recoupled 18
Regenerating 131
Regulate externalities 57
Reindeer
 husbandary 34, 35
Representative model 106
Resource
 consumption 140
Resource economy 129
Resource extraction 9, 141
Resource stock 181

Resource utilization 9
Ricardo, David 134
Royalty 138

S

Sea mammals 145
Seabed Authority 122
Sensitive dependence on initial conditions 87
Separating waste 3
Smith, Adam 22
Smoke abatement 60
Social accounting matrix 184
Social cost-benefit 168
Social fabrics 102
Sources of pollution 41
Sourcing 96-97
Stable systems 83
Standard National Accounts 174, 180
Steady-state equilibrium 107
Steel factory 59
Steel manufacturer 57
Stochastic control 125
Subsidiarity principle 12
Substitute commodity 207
Sulphur dioxide 24, 37
Surface and groundwater 187
Sustainable development 4, 18

T

Tatonnement process 20
Technical assumption 168
Threshold values 110
Toxic substance 31
Toxic waste 92
Transboundary pollution 80
Transferring resources 50
Transformation curve 48, 49
Transparent 216
Travel cost method 199
Tropical forests 212

U

Uncertainty 8, 42-45, 112-125
  case studies 117-121
  Law of the sea 121-123
  Modelling 112, 115
  single equations 115-117
  sources 112, 123-125
Undepletable externalities 24
Unidirectional externalities 57-59

User costs 138
Utilization 140

V

Value environmental
assets 196-199

W

Waste disposal 110
Waste management 16
Welfare
economic 190
Well-being 167, 168

●●●